Genetic Engineering 3

Genetic Engineering 3

Edited by

Robert Williamson

Professor of Biochemistry,
St Mary's Hospital Medical School,
University of London

ACADEMIC PRESS · 1982

A Subsidiary of Harcourt Brace Jovanovich, Publishers

London · New York · Paris · San Diego
San Francisco · São Paulo · Sydney · Tokyo · Toronto

ACADEMIC PRESS INC. (LONDON) LTD
24/28 Oval Road,
London NW1

United States Edition published by
ACADEMIC PRESS INC.
111 Fifth Avenue,
New York, New York 10003

British Library Cataloguing in Publication Data

Genetic engineering. — Vol. 3
   1. Genetic engineering
   575.1   QH442

ISBN 978-1-4615-7080-6   ISBN 978-1-4615-7078-3 (eBook)
DOI 10.1007/978-1-4615-7078-3

LCCCN 80—41976

Contributors

W.J. Brammar   *Department of Biochemistry, University of Leicester, Leicester LE1 7RH, UK*

K.E. Davies   *Biochemistry Department, St Mary's Hospital Medical School, University of London, London W2, UK*

P.W.J. Rigby   *Cancer Research Campaign, Eukaryotic Molecular Genetics Research Group, Department of Biochemistry, Imperial College of Science and Technology, London SW7 2AZ, UK*

R. Thompson   *Institute of Virology, University of Glasgow, Church Street, Glasgow G11 5JR, UK*

Preface

Like many genetic engineers, I have recently been receiving the attention of various venture capital companies, international drug houses and Members of Parliament. I will not discuss which of these approaches are most welcome, but it did cause me to consider the speed of advance in genetic engineering, and the implications of this rapid growth. There were few who anticipated it — only five years ago, most scientists thought applications would come at the end of the century, yet we see products such as insulin and interferon already available for clinical testing.

In Europe in general and Britain in particular, this explosive growth in our own field has coincided with a general industrial depression and a marked reduction in funding for biomedical research. The brain drain from Britain is a serious matter, for we are losing the best of our younger scientists, on whom we would rely to train the next generation of molecular biologists. These volumes have come from British labs (mostly because I happen to be based in London, and my contacts and friends are here), and I feel that the quality of the contributions also shows that our current research is of a high standard. It has been based on a scientific and social philosophy of service to both the scientific and the wider community. I hope that the entry of big money into the field will not distort this philosophy, nor in the end destroy the ethics of the scientific community that has given birth to the new genetics.

This volume continues the policy of its predecessors. There are three major articles, from Russell Thompson on plasmid vectors, Bill Brammar on phage vectors and Peter Rigby on the expression of cloned genes in eukaryotic cells using vectors based on viruses and similar systems. In addition, Kay Davies has prepared a list of all recombinants containing eukaryotic genes, as of October 1981. We hope to update this occasionally, although I expect that computerized lists available on line via satellites will eventually supplant it.

Those who have read the first two volumes of this series seem to enjoy them and use them — I hope the same will be true for this effort. When Volume 4 joins the series, the set will represent a complete description of the state of the art of genetic engineering at this time, suitable for those learning about the field, entering it for the first time, or working in it actively.

London, 29 November 1981   *Bob Williamson*

Contents

Plasmid and phage M13 cloning vectors

R. Thompson

Vectors based on bacteriophage lambda

W.J. Brammar

Expression of cloned genes in eukaryotic cells using vector systems derived from viral replicons

P.W.J. Rigby

A comprehensive list of cloned eukaryotic genes

K. E. Davies

Plasmid and phage M13 cloning vectors

RUSSELL THOMPSON

Institute of Virology, University of Glasgow, Glasgow, UK

1

I   Introduction

The essence of DNA cloning is the joining of a stretch of DNA of interest to a vector molecule which serves to propagate that DNA segment in bacteria. Vectors for *Escherichia coli* are derived from the natural phages and plasmids of this organism. Plasmid vectors have figured in all of the early achievements of recombinant DNA technology, the cloning and synthesis in *E.coli* of insulin and other hormones, interferon and animal virus antigens. To a large extent this reflects the wide range and versatility of the plasmids available for DNA cloning, a range that is continually expanding. The purpose of this chapter is to review the plasmid vectors currently available for cloning DNA in *E.coli* cells and to provide sufficient background material to enable the reader to follow the constant improvements in vector systems. Considerable effort has been directed at making *E.coli* cells an efficient source of the protein encoded by the cloned segment; vectors designed to allow expression of cloned genes will be discussed in detail in a separate chapter (see Carey, this series, Vol. 4) and they will be dealt with fairly lightly here. Plasmids which expand the cloning range to host cells other than *E. coli* will be discussed and finally, the special advantages of using the single-stranded DNA phages, such as M13, as cloning vectors will be described.

The application of DNA cloning techniques to the study of plasmids themselves has led to a considerable increase in our understanding of several fundamental plasmid properties such as replication, partition and copy number control. To provide a background for the more detailed discussion of plasmid vectors we will first turn to these more general topics.

II   Bacterial plasmids

A   Plasmid genes

Plasmids are extrachromosomal, self-replicating and stably inherited nucleic acid molecules. All plasmids so far isolated from bacteria have been molecules of double stranded circular DNA. Their stable inheritance suggests that plasmids may code for functions involved in their *replication* and *segregation* into daughter cells at cell division. Indeed genes involved in these two processes would seem to be the sole requirement for a piece of DNA to exist in the plasmid state and cryptic plasmids, which have no detectable phenotypic effect on their host cells, have been found (Kretschmer *et al.*, 1975). The vast majority of plasmids however carry many more genes than the minimum required for maintenance within a bacterial cell. Some of these genes confer on the host cell properties of medical and economic importance such as antibiotic or heavy metal resistance, virulence, toxin production and the ability to degrade exotic organic compounds. An extensive discussion of these properties, as well as much other background information can be found in the books of Falkow (1975) and Broda (1979).

Naturally occurring plasmids have been modified by *in vivo* and *in vitro* genetic manipulations to improve their usefulness as vectors for DNA cloning. The replication and transfer properties of plasmids are central to these improvements.

B   Plasmid replication in *E.coli*

Three aspects of plasmid replication which relate to their use as cloning vectors are:
(a) the number of plasmid copies per chromosome
(b) the size of the region essential for replication and partition
(c) the phenomenon of plasmid incompatibility

1   *Copy number*

Plasmids are maintained at characteristic copy numbers relative to the host chromosome. There is a continuous spectrum of copy numbers but it is convenient to define two groups: the low copy number plasmids present at a level of 1–5 copies per chromosome and the multicopy plasmids present at 15 or more copies per chromosome. Plasmid mutants which affect the copy number have been isolated from both classes showing the involvment of plasmid functions in copy number control. Gustafsson and Nordström (1978) isolated temperature sensitive and amber copy mutants of the low copy number drug resistance plasmid R1–19. The mutants have an elevated copy number under non-permissive conditions and exhibit an increase in their level of ampicillin resistance corresponding to the increased gene dosage of the β-lactamase gene on the plasmid. A second type of copy mutant, isolated by Nordström's group is a thermosensitive runaway replication mutant. Uhlin *et al.* (1979) have made derivatives of these mutants into useful cloning vectors which after a temperature shift replicate rapidly such that within a short period the plasmid DNA represents 75% of the total DNA. This runaway replication is lethal to the cell.

Copy mutants have also been isolated from the derivatives of the multicopy plasmid ColE1, a replicon from which several useful vectors have been derived. Gelfand *et al.* (1978) isolated a copy mutant which was present at the level of 30% of total intracellular DNA compared to 5% for the parental plasmid. A spontaneous deletion derivative of the mutant was still maintained at a level of 30% of the total DNA and since the plasmid DNA was smaller the copy number of the deletion mutant must have increased to maintain the same plasmid DNA level. This observation led Gelfand *et al.* (1978) to suggest that copy number is regulated by a plasmid specific factor that represses replication. The characteristic copy number is determined perhaps by the affinity of a repressor for its binding site, as had been suggested by others (Pritchard *et al.*, 1969; Cabello *et al.*, 1976). In the absence of this repressor, copy number would increase until host-encoded functions required for replication, such as DNA polymerase, became limiting. Direct evidence for a ColE1 coded replication repressor such as the isolation of an amber suppressible copy mutant has so far not been obtained. But consistent with the idea of negative control of ColE1 copy number is the finding by Shepherd *et al.* (1979) that the ColE1 elevated copy number mutant is recessive and falls when a wild-type ColE1 plasmid is present in the same cell. They have mapped a 2 kb region spanning the replication origin and presumably coding for the replication repressor which can suppress the DNA overproducer phenotype of the copy mutant in *cis* or in *trans*.

Johnson and Willetts (1980) have reported a bacterial strain which can stably accommodate 39% of its total DNA as plasmid. Thus the limiting level of plasmid DNA in a viable cell may be 30—40% of the total DNA. The manipulation of copy number allows manipulation of the dosage of cloned genes; the effect of this on the expression of the gene products will be discussed in a later section.

A finding of relevance to those interested in growing plasmid-containing cultures on an industrial scale is that copy number can fall during nutrient-limited growth. Jones *et al.* (1980) grew a ColE1-containing strain in a chemostat under conditions of glucose or phosphate limitation and found that during 80 generations the plasmid content of the cells fell five-fold. The ColE1 plasmid contained a transposon Tn*1* insertion coding for ampicillin resistance. Subculture in media containing ampicillin could reverse the drop in copy number showing that the decrease was not due to selection of mutants with a lower copy number but rather was a phenotypic change in response to nutrient limitation.

2   The minimal replicon

Small size is a desirable feature in a cloning vector. It maximizes the ratio of passenger to vector DNA and simplifies restriction digest patterns such that mapping the cloned segment and isolating fragments for sequencing are easier. These considerations lead to the question of what is the minimum component of a plasmid that can direct its own replication. Many cloning vectors have been derived from the small multicopy plasmids ColE1, pMB1 and P15A, which share the properties of continued replication in the absence of protein synthesis and dependence on DNA polymerase I (for reviews of plasmid replication see Kolter and Helinski (1979) and Staudenbauer (1978)). A 580 bp fragment from the replication origin region of pMB1 contains all of the genetic information necessary for replication as a plasmid in *E.coli* cells (Backman *et al.*, 1978). The transition point from primer RNA to DNA has been mapped for the closely related plasmid ColE1 (Bird and Tomizawa, 1978). It is 13 bp from one end of the 580 bp fragment showing that no information is necessary downstream from the origin. At the other end of this fragment is a region which is transcribed *in vivo* to yield a 100-nucleotide transcript. Backman *et al.* (1978) have proposed a nomadic primer model in which this transcript is processed and migrates to the replication origin where it can act as a primer for DNA synthesis. However, Oka *et al.* (1979) have isolated ColE1 derivatives which lack the nomadic primer region so that the source of the primer RNA remains obscure. The 580 bp fragment, which has been completely sequenced, contains no obvious sequence which might code for a polypeptide. This, together with experiments which indicate that no ColE1 encoded protein is needed for replication (Donoghue and Sharpe, 1978; Kahn and Helinski, 1978) suggests that the plasmid supplies a stretch of DNA which is recognized as a replication origin and that host enzymes are solely responsible for the replication reactions.

Such a simple picture is not the case for all plasmids. The antibiotic resistance plasmid R6K is 38 kb in size and a 2 kb segment from it is stably maintained in *E.coli* at the same copy number as the parental plasmid. The 2 kb fragment contains a gene *pir* coding for the π protein, which is essential for R6K replication, and an *ori* region which functions as an origin of replication (Kolter *et al.*, 1978). Kolter and Helinski (1979) have proposed a model in which the π protein has a dual role firstly as a positive element regulating the frequency of initiation of replication of the R6K origin and secondly as a negative element regulating its own synthesis. The π protein is proposed to regulate its own synthesis by binding to the nucleotide sequence repeats in the operator region of the *pir* gene and repressing transcription.

Replication alone is not enough to ensure stable inheritance of plasmid DNA molecules within a growing bacterial population. The plasmid molecules must be segregated accurately into the daughter cells at cell division. To accomplish this, plasmids have a stretch of DNA which is functionally equivalent to the centromere of eukaryotic chromosomes. This insight has come from recent experiments of Meacock and Cohen (1980). They have identified a locus, designated *par* for partition, that is required for stable plasmid maintenance. The *par* locus of pSC101 lies in a 270 bp segment adjacent to the replication origin but is not directly associated with plasmid replication functions. Partition defective plasmids which lack a *par* locus can be maintained in a population by continuous selection, for instance for expression of a plasmid coded drug resistance gene. On removing the selection, however, the plasmid is slowly lost from the growing population and the rate of segregation of plasmid-free cells is proportional to the plasmid copy number.

Meacock and Cohen found that in the course of the DNA manipulations used to convert the naturally occurring plasmid P15A into the cloning vector plasmid pACYC184 the *par* locus of P15A had been deleted. As a result of this pACYC184 is unstable and is lost from cells cultured for long periods in non-selective medium. Cloning of the DNA fragment containing the *par* locus of pSC101 into pACYC184 can restore plasmid stability; the pSC101 *par* locus can function actively to segregate the unrelated plasmid pACYC184. This stabilization of a *par*⁻ plasmid only works in the *cis* configuration, that is a *par* locus is only active in segregation of the DNA molecule to which it is physically linked. The *par* locus is presumably a DNA site which interacts with cellular components to accomplish partitioning of plasmid DNA molecules during cell division.

In seeking to reduce the size of vector plasmids, DNA regions which were non-essential for replication and which did not contain selectable markers have been removed. As a consequence of this many cloning vectors probably are *par*⁻. Jones *et al.* (1980) found that pBR322 and pMB9-containing cells give rise to plasmid-free segregants after about 30 generations of growth in a nutrient limited chemostat. This rate of loss is likely to be due to two factors, the lack of accurate partitioning and the drop in plasmid copy number in conditions of nutrient limitation. While the absence of a *par* locus from vector plasmids will not affect their growth on a laboratory scale, particularly if care is taken that the inoculum is 100% plasmid-carrying, it may lead to problems on an industrial scale. The problem can be overcome by simply cloning the 270 bp *par* fragment from pSC101 into the vector. Alternatively a continuous selection could be applied by incorporating an essential host gene, say a cell wall gene, into the vector and using a host that was deleted for that gene.

3   Plasmid incompatibility

It is possible to isolate cells containing any number of different plasmid types provided that the plasmids are from different incompatibility groups. Two plasmids which cannot be stably maintained in the same cell are said to be incompatible; they are members of the same incompatibility group. Naturally occurring plasmids have been found to fall into a large number of incompatibility groups (see Appendix B in Bukhari *et al.*, 1977). The many cloning vectors have been derived from a small number of parental plasmids. Of these ColE1 and pMB1 fall into the same incompatibility group so that experiments to examine the interaction of the products of genes cloned on different derivatives of these two plasmids are not practical. A third small multicopy plasmid P15A which is the progenitor of several vector plasmids is, however, compatible with ColE1 and pMB1 (Chang and Cohen, 1978). Likewise the plasmids pSC101, F and RP4 all fall into different incompatibility groups such that vectors derived from one of these plasmids are stable in cells containing plasmids derived from any of the others.

C   Transfer and mobilization

Plasmid DNA molecules range from 2 kb to over 200 kb in size; this range is similar to that of organelle and viral genomes. Those larger than about 30 kb often carry a set of genes which mediate conjugal transfer of the plasmid DNA to other bacterial cells. The best studied and archetypal example of these conjugative plasmids is the F factor of *E.coli*, although many other plasmid transfer systems have been described. Interbacterial DNA transfer by conjugation is a complex process requiring the products of at least 20 transfer genes (for review see Clark and Warren, 1979; Willetts and Skurray, 1980). Plasmids that are too small to code for complete transfer systems can often be transferred if a conjugative plasmid is present in the same cell. This process of transfer of a small, non-conjugative plasmid by a coresident, large, conjugative one is termed mobilization.

Plasmid mobilization has been extensively studied using the small plasmid ColE1. It has been shown that mobilization requires both a specific site on the ColE1 DNA and ColE1-specified diffusible gene products (Warren *et al*, 1978). About one-third of the ColE1 genome or 2 kb of DNA is necessary for mobility and mobilization deficient mutants of ColE1 have been grouped into three complementation groups (Dougan *et al.*, 1978; Inselburg and Ware, 1979). Most cloning vectors derived from ColE1 or the related plasmid pMB1 have lost the DNA region which codes for the mobility proteins.

The proteins can, however, be supplied in *trans* by a compatible plasmid such as ColK. The mobility proteins probably act at a site designated *nic* (see Clark and Warren, 1979, for a discussion of this point). The ColE1 *nic* site has been sequenced (Bastia, 1978) and this sequence is conserved in pBR322 (Sutcliffe, 1978a). Several vectors have had the *nic* site deleted during their construction (see below). They cannot be mobilized and the only possible route of conjugal transmission for such plasmids is if they physically become part of a conjugative plasmid by recombination to form a fused or cointegrate plasmid. The use of recombination-deficient host strains such as *recA* strains removes this possibility, so that vectors deleted for *nic* in a *recA* host are considered more biologically "contained" than *nic*⁺ vectors.

III   General purpose amplifiable vectors

To be of use as a cloning vector a plasmid must have a unique site for one or more restriction enzymes at which insertion of DNA does not interfere with plasmid replication functions. New plasmids formed by inserting DNA fragments at these restriction sites must be capable of reintroduction into bacterial cells and cells inheriting them should be easily identifiable. This latter point usually means that insertion of new DNA at a particular restriction site must leave at least one selective marker on the plasmid intact. The first DNA cloning experiments to be carried out (Cohen *et al.*, 1973) used pSC101, a plasmid isolated from *Salmonella* (Cohen and Chang, 1977). This plasmid contains a single *Eco*RI site in a position such that cloning of DNA into this site does not affect either replication or the only marker selective for the presence of the plasmid, a gene coding for resistance to tetracycline. Since then, the trend has been to develop plasmids of minimal size that carry two or three selective markers. Often the unique cloning sites are within one or other of the selective markers such that insertion of DNA at the site inactivates the particular marker and allows ready identification of plasmids carrying DNA inserts. The early cloning vectors are described in the review by Collins (1977). More recent reviews are those by Brammar (1979), Sherratt (1979), Bolivar and Backman (1980), Bernard and Helsinki (1980), Kahn *et al.* (1979) and Timmis (1981).

A   Choice of vector

Often the sole purpose of cloning a DNA fragment is to allow isolation of large quantities of the DNA in pure form. Plasmids derived from

ColE1, pMB1 or P15A are particularly useful for this purpose for several reasons. They are multicopy plasmids maintained in cells at levels of 10 or more copies per chromosome equivalent. The copy number can be amplified to as much as 1500 by treatment of the culture with inhibitors of protein synthesis such as chloramphenicol or spectinomycin (Clewell, 1972; Chang and Cohen, 1978). This allows the isolation of the plasmid DNA with yields in excess of 1 mg/litre of cells. A further 2—3-fold amplification may be achieved by addition of high concentrations of uridine (Norgard *et al.*, 1979). The P15A based vectors are however much less amplifiable than the others (Chang and Cohen, 1978). As discussed above, only a small segment of these plasmids is necessary for replication; regions outside of this can therefore be deleted during rearrangement of the cloned segment.

Table 1 lists the most useful of the multicopy vectors and physical maps are presented in Fig. 1. Most of the plasmids shown in Fig. 1 contain the tetracycline resistance gene originally found on pSC101 and the maps have been aligned using the common *Hin*dIII site in this gene. The plasmid pKC7 was constructed by replacing the small *Hin*dIII—*Bam*HI fragment of pBR322 with a fragment containing the kanamycin resistance gene of transposon Tn5 (Rao and Rogers, 1979). The position of the *Hin*dIII site is therefore unchanged. The only plasmid lacking the *Hin*dIII site, pACYC177, has been arbitrarily linearized from one end of the Ap^r gene.

In general it is desirable to be able to recover the cloned fragment free of vector DNA by restriction enzyme cleavage. Choice of vector is to some extent influenced by the enzyme(s) used to generate the desired fragment (but see the section on regeneration of restriction sites). If the fragment to be cloned encodes a function which can be selected, then all of the vectors with unique sites for the appropriate enzyme would be equally useful. This is rarely the case and vectors may be preferred which have restriction sites positioned such that insertion of a DNA fragment inactivates a particular gene. For example, cloning *Eco*RI fragments into pBR322 does not give a detectable change in plasmid phenotype. Thus, to distinguish within a transformed cell population between cells carrying the vector alone and cells carrying vector plus insert, properties such as the size of the plasmid DNA or the potential to hybridize to a suitable probe must be examined. In contrast, cloning into the *Eco*RI site of pBR328 inactivates the chloramphenicol resistance gene, and screening for chloramphenicol sensitive transformants thus identifies a population which is greatly enriched for plasmids carrying inserts (there will be a background of Cm^s clones with no insert which arise by aberrant recircularization of the vector).

Table 1 General purpose plasmid vectors.

Plasmid*	Size (kb)	Genetic markers	Cloning sites/phenotype†	Reference
pMB9	5.4	Colicin immunity Tc^r	*Eco*RI, *Sma*I, *Hpa*I/None *Hind*III, *Bam*HI, *Sal*I,/Tc^s	1
pBR322	4.4	Tc^r, Ap^r	*Eco*RI, *Bal*I, *Ava*I, *Pvu*II/None *Hind*III‡, *Bam*HI, *Cla*I, *Sal*I, *Sph*I/Tc^s *Pst*I, *Pvu*I/Ap^s	2
pAT153	3.7	Tc^r, Ap^r	As in pBR322 except *Pvu*II absent	3
pBR324	8.2	Colicin immunity Colicin production Tc^r, Ap^r	*Eco*RI, *Sma*I/Cea^- *Hind*III, *Sal*I, *Bam*HI/Tc^s	4
pBR325	5.4	Tc^r, Ap^r, Cm^r	*Eco*RI/Cm^s *Pst*I, *Pvu*I/Ap^s *Hind*III, *Bam*HI, *Sal*I, *Sph*I/Tc^s *Ava*I/none	4
pBR327	3.3	Tc^r, Ap^r	*Hind*III, *Bam*HI, *Sal*I, *Sph*I/Tc^s *Pst*I, *Pvu*I/Ap^s *Eco*RI, *Ava*I/none	5
pBR328	4.9	Tc^r, Ap^r, Cm^r	*Hind*III, *Bam*HI, *Sal*I, *Sph*I/Tc^s *Pst*I, *Pvu*I/Ap^s *Eco*RI, *Bal*I, *Pvu*II/Cm^s *Ava*I/none	5
pKC7	5.8	Ap^r, Km^r	*Bgl*II, ‡ *Bcl*II/Km^s *Pvu*I/Ap^s *Eco*RI, *Hind*III, *Sma*I, *Xho*I, *Bst*EII, *Bam*HI/none	6

pMK2004	5.2	Tc^r, Ap^r, Km^r	*Bam*HI, *Sal*I/Tc^s *Sma*I, *Xho*I/Km^s *Pst*I/Ap^s *Eco*RI/none	7
pACYC177	3.7	Ap^r, Km^r	*Pst*I, *Hinc*II/Ap^s *Hind*III, *Sma*I, *Xho*I/Km^s *Bam*Hl/none	8
pACYC184	4.0	Tc^r, Cm^r	*Hind*III, *Bam*HI, *Sal*I/Tc^s *Eco*RI/Cm^s	8
pMK16	4.5	Tc^r, Km^r Colicin immunity	*Bam*HI, *Sal*I, *Hinc*II/Tc^s *Xho*I, *Sma*I/Km^s *Eco*RI/none	7

*The plasmids pMB9 to pMK2004 are based on the pMB1 replicon. pACYC177 and 184 are based on the p15A replicon and pMK16 is based on the ColE1 replicon.

†Unique restriction sites available for cloning are listed. Insertion of DNA into sites grouped on the same line results in the same change in plasmid phenotype (indicated beyond the oblique stroke). The superscripts s and r indicate sensitivity or resistance to the following antibiotics: Tc, tetracycline, Ap, ampicillin, Cm, chloramphenicol, Km, kanamycin. Cea⁻ indicates failure to produce colicin.

‡Note the discussion in the text about occasional failure of insertional inactivation when cloning into *Hind*III and *Bgl*II sites.
1, Bolivar *et al.* (1977a); 2, Bolivar *et al.* (1977b); 3, Twigg and Sherratt (1980); 4, Bolivar (1978); 5, Soberon *et al.* (1980); 6, Rao and Rogers (1979); 7, Kahn *et al.* (1979); 8, Chang and Cohen (1978).

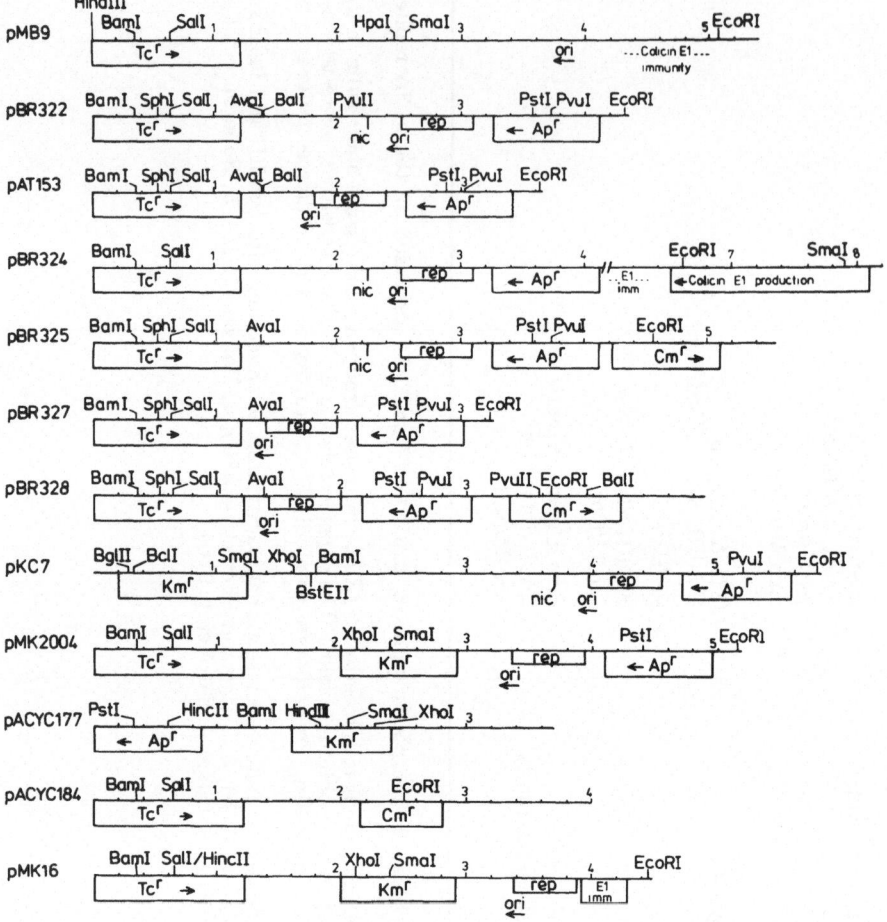

Figure 1 Physical maps of plasmid vectors showing unique restriction sites available for cloning. The maps are divided into units of 0.2 kb and they are aligned (with one exception) at the *Hind*III site at the beginning of the Tc^r gene. The exception is pACYC177 which has been linearized from one end of the Ap^r gene. The direction of transcription of the antibiotic resistance genes, where known, is indicated by an arrow. The region required for plasmid replication and the origin and direction of replication as well as the *nic* site needed for mobilization are shown. There is a *Cla*I site immediately preceding the *Hind*III site in the Tc^r gene; this is a unique site in pBR322, pAT153 and pBR327, but whether it is also unique in the other vectors is not known. Note that pMK2004 and pMK16 have a second *Hind*III site in the Km^r gene and thus there is not a unique *Hind*III cloning site in these plasmids. The recognition sequence for *Sph*I is GCATG^{v}C (Fuchs *et al.*, 1980). Novick *et al.* (1976) have proposed a uniform plasmid nomenclature that takes the form pXY1234. See Table 1 for references.

Insertional inactivation of a plasmid marker, although useful, is not essential since other methods can generate populations in which almost all of the cells contain recombinant plasmids rather than the vector alone. Treatment of vector DNA with phosphatase after restriction enzyme cleavage removes the terminal phosphate groups and prevents rejoining of the vector unless a fragment of passenger DNA is present (Ullrich *et al.*, 1977). DNA joining by means of annealing complementary homopolymer extensions added to vector and passenger DNA by terminal transferase can also ensure that most plasmids recovered carry inserts (Jackson *et al.*, 1972).

When cloning fragments with ends generated by different restriction enzymes a vector must be chosen which will accommodate the fragment while still leaving one selective marker intact. A fragment with one *Sal*I end and one *Pst*I end could be cloned into pMK2004 and leave the Km^r gene intact. Consideration should also be given to the suitability of the vector for further *in vitro* manipulation of the cloned fragment such as the deletion of defined segments by making use of restriction sites flanking the insert.

Vectors derived from ColE1 and pMB1 share extensive DNA homology due to the sequence relatedness of the parental plasmids (Betlach *et al.*, 1976) and the common Tc^r gene. Similarly, p15A is 60% homologous to ColE1 (Goebel and Schrempf, 1972). All the vectors listed in Table 1 therefore have some degree of sequence homology. This common background must be eliminated in order to use DNA from a particular clone as a hybridization probe to screen a library of clones for those with the same insert (Maniatis *et al.*, 1978). Likewise, the "chromosome walking" technique, which identifies the genomic fragments flanking a cloned segment by screening a library for clones with inserts that overlap that segment, demands that background vector hybridization be removed (Chinault and Carbon, 1979). Vector and passenger DNA can be separated by restriction enzyme cleavage followed by gel electrophoresis. Alternatively, the DNA can be labelled by nick translation, denatured in the presence of an excess of vector DNA and renatured to a low C_0t value. The single stranded passenger DNA can then be separated from the double stranded vector DNA by hydroxylapatite chromatography (Chinault and Carbon, 1979). A third possibility involves the use of the plasmid RSF1030 (Heffron *et al.*, 1975). This 8.3 kb ampicillin resistant plasmid has a unique *Bam*Hl site and does not hybridize to either ColE1- or pMB1-derived plasmids (Morrow, 1979). Cloning into RSF1030 would remove the need to separate vector and passenger DNA when screening, for example, a pBR322-based library.

Proteins encoded by the cloned segment and synthesized in *E.coli* can be identified by either the minicell or the maxicell technique. Minicells are small, spherical, chromosomeless cells produced by aberrant cell divisions (see Frazer and Curtiss, 1975 for review). Plasmid DNAs segregate into minicells and purified minicell preparations allow the labelling of plasmid encoded proteins in the absence of any background from chromosomal proteins. The maxicell technique (Sancar *et al.*, 1979) employs a UV-repair deficient strain which degrades UV-damaged DNA. Because of the size difference between plasmid and chromosome, a UV dose can be selected which causes degradation of the chromosomal DNA while leaving the plasmid intact and allowing labelling of plasmid proteins in the absence of any chromosomal background. Roberts *et al.* (1979a) used the maxicell technique to show synthesis of SV40 t antigen in *E.coli*. Proteins from both vector and passenger DNAs are displayed by these techniques. The plasmid pHA105 is noteworthy here since this 2.4 kb plasmid contains a single *Eco*RI site and does not encode any polypeptides larger than 16 000 mol. wt (Avni and Markovitz, 1979).

B   Choice of host cell

The guidelines governing recombinant DNA work have been evolving faster than the vector systems to which they relate. The choice of an *E.coli* strain as a cloning host is in many instances affected by these guidelines. The disabled host strain χ1776, developed by Curtiss *et al.* (1977) has now been supplemented with a series of strains, MRC1, 7, 8 and 9 constructed by Brenner (1979). The *recA* disabled strain MRC8 has been designated an especially disabled system when used in conjunction with non-mobilizable vectors.

Many experiments no longer need a disabled host and often a *recA* host with a non-mobilizable vector will suffice. Two strains have been widely used in this role, HB101 and DH1 (Boyer and Roulland-Dussoix, 1969; D. Hanahan, pers. comm.). Both are *recA* and restriction deficient, transform well and give high yields of plasmid DNA, but since transformation efficiencies vary between laboratories there may be local preferences.

C   Selective markers

Most vectors use antibiotic resistance as a selectable marker, although metabolic markers and immunity to colicin E1 may also be used (Hershfield *et al.*, 1974). A recent advance has been the development of a positive selection system for the loss of tetracycline resistance (Bochner *et al.*, 1980). This technique is based on the observation that tetracycline resistant cells are hypersensitive to lipophilic chelating agents, such as fusaric acid, allowing the formulation of a medium on which only tetracycline sensitive and not resistant cells grow. When cloning into a site which gives insertional inactivation of a Tc^r gene, use of the fusaric acid medium will yield a population of colonies all of which potentially contain passenger DNA. One drawback of this technique is that it is strain dependent and does not work for C600 or HB101, two commonly used hosts for cloning experiments. However a modified formulation has been devised for these and other *E.coli* strains (Maloy and Nunn, 1981).

Although the Tc^r gene of pMB9 and the pBR322 series of plasmids originally came from pSC101, the *in vitro* manipulations during the construction of these plasmids have caused some changes. Tetracycline resistance is no longer inducible in pBR322 as it is in pSC101 (Tait *et al.*, 1977, Tait and Boyer, 1978a, b). Resistance to tetracycline is due to a permeability effect (Bochner *et al.*, 1980) and although the number of proteins involved is still uncertain, it seems clear that the *Hin*dIII site lies in the promoter for the Tc^r gene(s) (Sutcliffe, 1978; Rodriguez *et al.*, 1979). Not all insertions at the *Hin*dIII site result in a Tc^s phenotype; if the inserted fragment contains a promoter directed into the Tc^r gene then tetracycline resistance may be expressed (Widera *et al.*, 1978). A similar situation is seen with the plasmid pKC7, where not all insertions at the *Bgl*II site cause loss of kanamycin resistance (Rao and Rogers, 1979).

Enrichment for tetracycline- or chloramphenicol-sensitive clones can be achieved by making use of the fact that these two antibiotics are bacteriostatic. In a mixed population of Tc^r and Tc^s cells the Tc^r cells will continue to grow in the presence of tetracycline and can be killed by D-cycloserine which is lethal only for growing cells (Bolivar *et al.*, 1977a). Tc^s cells can then be recovered after removal of the antibiotics.

D   Regeneration of restriction sites

The repertoire of cloning vectors is incomplete in the sense that cloning sites are not available for every known restriction enzyme specificity. Considerable flexibility can be obtained with presently available cloning sites by adopting the approach of Backman *et al.*, (1976). They exploited the fact that *Eco*RI produces staggered ends with a single-stranded 5′ extension. These ends can be converted to flush ends by treatment with DNA polymerase I and deoxyribonucleoside triphosphates since the recessed 3′ end can be extended by copying the protruding 5′ end. This reaction generates a flush end with the sequence 5′AATTC and ligation to a flush-ended fragment which has a 5′C (and hence a 3′G) will restore the *Eco*RI recognition sequence 5′GAATTC. Backman *et al.* (1976) used a fragment generated by *Hae*III which cuts 5′GG^vCC to produce a blunt end with a 5′C. Several other enzymes, *Alu*I, *Fnu*DII, *Pvu*II and *Bal*I produce a 5′C blunt end. In addition, the 5′ extensions produced by *Xba*I, *Xma*I, *Cla*I, *Eco*RII, *Hpa*II and *Taq*I can be filled in with DNA polymerase to yield 5′C blunt ends (Roberts, 1978). Because the recognition sequences of *Eco*RI, *Bam*Hl and *Sal*I all end with a 3′C, termini produced by all three enzymes can be filled in a ligated to a 5′C blunt end to regenerate the site. This technique, together with the use of oligodeoxyribonucleotide linkers containing restriction sites (Bahl *et al.*, 1976) greatly expands the usefulness of available cloning sites.

E   The pBR322 series

Vectors specifically tailored to the needs of DNA cloning have been developed from natural plasmids. This is neatly illustrated by examining the series of *in vivo* and *in vitro* steps leading to the construction of pBR322, one of the most widely used cloning vectors (see Bolivar, 1979, for review). The parental plasmid for this series is pMB1 and initial interest in this plasmid stems from the fact that it carries the genes coding for the *Eco*RI endonuclease and the corresponding methylase (Betlach *et al.*, 1976). Otherwise pMB1 is very like another naturally occurring plasmid ColE1. It carries genes coding for the antibiotic protein colicin E1 and immunity to the killing action of this protein, as well as genes and DNA sites needed for replication, partition and mobilization. The DNA of pMB1 is 8.3 kb in size.

The first step in the construction of pBR322 was the *in vivo* transposition of the ampicillin resistance transposon Tn*3* into pMB1 to yield the 12.1 kb plasmid pMB3 (Betlach *et al.*, 1976). This plasmid was reduced in size by digestion with *Eco*RI*, which recognizes the central tetranucleotide of the *Eco*RI recognition sequence (Polisky *et al.*, 1975), followed by religation. This process caused the loss of the ampicillin resistance gene but generated the 2.6 kb plasmid pMB8 which still carried the colicin immunity gene and had a unique *Eco*RI site (Rodriguez, *et al.*, 1976). DNA of pMB8 which had been linearized with *Eco*RI was then ligated to an *Eco*RI* digest of pSC101 DNA. Cells were transformed with the products of this ligation reaction and a plasmid designated pMB9 was isolated which carried genes for tetracycline resistance and colicin immunity and which contained a single *Eco*RI site.

A map of pMB9 is shown in Fig. 1. It is 5.4 kb in size and contains unique sites for six restriction enzymes. Since colicin immunity is not a particularly good selective marker, an ampicillin resistance gene was introduced by transposing Tn*3* onto pMB9 to give the plasmid pBR312. The size of pBR312 was reduced from 10.1 kb to 8.7 kb by partial *Eco*RI* digestion and religation to yield pBR313. As a consequence of the ability of *Eco*RI* digestion/ligation reactions to remove *Eco*RI sites or generate new ones, the *Eco*RI and *Hin*dIII sites of pBR313 are 31 bp apart as opposed to the 330 bp which separates them in pMB9. A further consequence of the size reduction is that the Amp^r gene in pBR313 can no longer be transposed (Bolivar *et al.*, 1977a). A second size reduction was achieved by isolating two separate fragments of pBR313 and rejoining them *in vitro* to yield pBR322 (Bolivar *et al.*, 1977b). The key features in the development of pBR322 have been addition of selectable drug resistance markers by either *in vivo* transposition or *in vitro* recombination together with size reduction to remove non-essential DNA segments.

The entire 4362-nucleotide-pair sequence of pBR322 has been determined (Sutcliffe, 1978a). This information is useful when constructing restriction maps of fragments cloned into this vector and makes pBR322 a convenient source of DNA fragments for use as size markers (Sutcliffe, 1978b). It has been noted from the sequence that two of the boundaries between DNA from different sources show no trace of the restriction sites which should be there, given the description of the construction. Sequences may have been lost by *in vivo* recombination events near unligated DNA termini during one or more of the construction steps (Sutcliffe, 1978a; Bolivar *et al.*, 1977). Despite the uncertainty over some of the stages in its construction, pBR322 remains a very useful cloning vector and improved derivatives continue to be made. Deletion of the *Hae*II B and H fragments by partial digestion and religation generates the plasmid pAT153 which is 0.17 kb smaller than pBR322, has a three-fold elevated copy number and has lost the *nic* site and thus cannot be mobilized by a conjugative plasmid (Twigg and Sherratt, 1980). pBR327, a similar deletion derivative of pBR322, has been isolated by Soberon *et al.* (1980) by partial *Eco*RII digestion. Insertional inactivation of vector markers by *Eco*RI fragments, a feature not offered by pBR322, can be achieved using derivatives containing genes for colicin E1 production or chloramphenicol resistance (see pBR324, pBR325 and pBR328 in Fig. 1; Bolivar, 1978; Soberon *et al.*, 1980). These improved derivatives mean that the pBR322 series will continue to be one of the most widely used plasmid cloning systems.

IV   Specialized vectors

A   Low copy number vectors

Vectors have been derived from plasmids like F and pSC101 which are naturally present at a low copy number. Such vectors allow the cloning of genes that might be lethal to the cell if present in many copies. Comparison of the effect of cloning the same gene on a low copy number or a multicopy vector will give insights into the effect of gene dosage on regulation of gene expression.

Low copy number vectors are listed in Table 2. A general proble is that their low copy number and small size makes isolation of lai amounts of the DNA difficult. This has been neatly circumvent in the case of pDF41 by joining this plasmid via its *Eco*RI site to i multicopy pMK16 (see Table 1). The resulting double replicon pDF42, is maintained at the high copy level of the pMK16 component and is amplifiable by chloramphenicol. Large amounts of pDF42 DNA can thus be prepared from which pDF41 can be recovered by *Eco*RI digestion (Kahn *et al.*, 1979).

B   Vectors designed to detect transcription control signals

An intact structural gene sequence devoid of its promoter and with a cloning site upstream provides a means of assaying cloned fragments for the presence of a promoter. West *et al.* (1979) have constructed several plasmids of this type of which pBRH1 is described in Table 2. They have used the ubiquitous tetracycline resistance gene which appears in most of the vectors discussed in the previous sections. The Tc^r gene of pBRH1 is not expressed because of insertion of a synthetic octanucleotide containing an *Eco*RI recognition sequence into the *Hin*dIII site in the Tc^r promoter. This gives rise to a "promotor probe" plasmid since cloning of *Eco*RI fragments which contain correctly oriented promoter will cause the Tc^r phenotype to ‑ expressed. The system has been used to examine which sequence elements of a promoter are important for *in vivo* activity (Rodriguez *et al.*, 1979; West and Rodriquez, 1980). DNAs from a variety of eukaryotic and prokaryotic sources have been screened for promoter activity (Neve *et al.*, 1979). Digests of *Euglena* chloroplast and *Saccharomyces cerevisiae* DNA yielded as many active promoter fragments as did *E.coli* DNA. Whether the regions on the eukaryotic fragments which are recognized by *E.coli* RNA polymerase are genuine eukaryotic promoters remains to be seen.

Similar "promoter probe" plasmids with cloning sites for a wider range of restriction enzymes have been constructed by An and Friesen (1979, see Table 2).

Casadaban and Cohen (1980) have described a series of plasmids which can be used to detect not only promoters but also transcription termination signals. These plasmids contain the structural genes of the *lac* operon fused to the inducible promoter from the arabinose operon. Between these two genetic elements lie cloning sites for *Hind*III, *Kpn*I, *Sma*I, *Hinc*II or *Hpa*I. In the native vector expression of the *lac* genes is inducible by arabinose. If a cloned fragment contains a correctly oriented promoter then *lac* expression becomes independent of the presence of arabinose. If the fragment contains a transcription ermination signal then *lac* will not be expressed, even in the presence of arabinose. Regulatory mechanisms acting on a cloned promoter as well as mutations which affect them can be conveniently assayed using the promoter probe plasmids.

C   Direct selection vectors

Direct selection for transformants containing recombinant plasmids considerably reduces the amount of screening needed to identify a desired clone. Selection for insertional inactivation of a tetracycline resistance gene has already been discussed. Four plasmid vectors allow direct selection (or screening) for clones containing inserts (Table 2).

Ozaki *et al.* (1980) have described a plasmid constructed by inserting the Ap^r transposon Tn*3* into ColE1 DNA. The transposon is inserted into the gene for colicin immunity. Cells containing this plasmid form minute colonies on plates containing mitomycin C since this compound induces the synthesis of colicin E1 and the cells lack immunity to the inhibitory effect of the colicin. Unique *Eco*RI and *Sma*I sites lie in the structural gene for colicin E1 (see the map of pBR324 in Fig. 1) and plasmids with DNA inserts cloned into these sites can no longer direct synthesis of the colicin. Cells containing such recombinan plasmids can form normal size colonies on mitomycin C plates.

Plasmid pKN80 carries a fragment of phage Mu DNA which encodes a killing function that is under the control of the prophage repressor (Schumann, 1979). This plasmid can be normally replicated in strains lysogenic for phage Mu but is lethal for non-lysogenic strains. Insertion of DNA at *Hpa*I or *Hind*II sites inactivates the killing function and the ability to yield Ap^r transformants of a strain which is not lysogenic for Mu affords a direct selection for clones with inserts at these sites.

A plasmid which allows visual screening for recombinant clones is

Table 2 Special purpose cloning vectors.

Plasmid	Replicon	Size	Genetic markers	Cloning sites/phenotype*	Remarks	Reference
Low copy number vectors						
pMF3	F	11.0	Ap^r	*Eco*RI, *Hin*dIII, *Bam*H1/none	1–2 copies/chromosome	1
pDF41	F	12.7	*trpE*$^+$	*Eco*RI, *Hin*dIII, *Bam*H1, *Sal*I/none	1–2 copies/chromosome	2
pRK2501	RK2	11.1	Tc^r, Km^r	*Sal*/Tc^s *Hin*dIII, *Xho*I/Km^s *Eco*RI, *Bgl*II/none	2–4 copies/chromosome	2
pHSG415	pSC101	7.1	Ap^r, Km^r, Cm^r	*Pst*I/Ap^s *Hin*dIII, *Xho*I/Km^s *Eco*RI/Cm^s *Bam*H1/none	5 copies/chromosome temperature-sensitive replication	3
Promoter probe vectors						
pBRH1	pMB1	7.8	Ap^r	Promoter containing *Eco*RI fragments result in expression of Tc^r when cloned in the correct orientation.		4
pGA39	p15A	4.6	Cm^r	*Hin*dIII, *Xma*I, *Pst*I or blunt ended fragments containing promoters result in expression of Tc^r when cloned in the correct orientation.		5
pGA46	p15A	4.4	Cm^r	As above for *Bgl*II, *Hin*dIII or *Pst*I fragments.		5
pMC81	ColE1	25.6	Ap^r	One of a series of plasmids with cloning sites between *lac* genes and *Para*, the promoter of the arabinose operon. Expression of *lac* in the absence of arabinose indicates a promoter on the cloned fragment.		6

				Cloning sites	Description	
Direct selection vectors						
pKY2289	ColE1	11.0	Ap^r		DNA insertion at *EcoRI* or *XmaI* site allows colonies to grow on plates containing mitomycin C.	7
pKN80	ColE1	15.8	Ap^r		DNA insertion at *HpaI* site inactivates a lethal gene and allows colony formation on Mu-non-lysogens.	8
pUR2	pMB1	2.6	Ap^r		Direct screening rather than direct selection. Plasmids with inserts in the unique *EcoRI*, *Bam*HI or *Hind*III sites yield white rather than blue colonies on X-gal plates.	9
pTR262	pMB1	5.0	λ immunity, *lac* operator		Insertion of *Hind*III or *Bcl*I fragments switches on a tetracycline resistance gene.	10
Cosmid for packaging in λ heads						
pHC79	pMB1	6.4	Ap^r, Tc^r	*Cla*I, *Hind*III, *Bam*HI, *Sal*I/Tc^s *Pst*I/Ap^s *Eco*RI, *Eca*I/none	Can be packaged *in vitro* into λ heads when linked to 30–40 kb fragments.	11

*Notation is as in Table 1.

1, Manis and Kline (1977); 2, Kahn *et al.* (1979); 3, Hashimoto-Gotoh *et al.* (1980); 4, West *et al.* (1979) 5, An and Friesen (1979); 6, Casadaban and Cohen (1980); 7, Ozaki *et al.* (1980); 8, Schumann (1979); 9, Rüther (1980); 10, Roberts *et al.* (1980); 11, Hohn and Collins (1980).

pUR2 (Rüther, 1980). On appropriate plates, colonies containing pUR2 alone are blue and those containing pUR2 with a DNA insert are white. The indicator system is the same as that used in the single-stranded phage vector M13mp2 and is described in more detail in the section on M13 vectors.

Plasmid pTR262 provides a unique positive selection system in which colonies that have acquired vector DNA with an insert of passenger DNA are tetracycline resistant (Roberts *et al.*, 1980). pTR262 was constructed by replacing the *Pst*I—*Eco*RI fragment of pBR322 with a fragment of phage λ DNA that contains the intact *c*I repressor gene as well as the rightward λ promoter, p_R. The promoter for the pBR322 Tet gene has been removed leaving the Tet gene dependent on transcription from p_R. Since this transcription is repressed by the *c*I protein the Tet gene is not expressed from native pTR262 DNA. Cells containing pTR262 are immune to λ infection by virtue of the presence of the *c*I protein. They are also *lac*-constitutive since the plasmid contains a copy of the *lac* operator and this titrates out the cellular *lac* repressor. Either of these phenotypes can be used to demonstrate the presence of pTR262. Passenger DNA inserted at the unique *Hin*dIII or *Bcl*I sites in the plasmid interrupts the coding region of the *c*I gene. Cells containing such plasmids synthesize no *c*I protein, transcription from p_R can therefore proceed and result in expression of the Tet gene. Plating on tetracycline directly identifies these clones. The direct selection for clones with inserts afforded by vectors such as these is particularly useful in the more technically demanding protocols such as cDNA cloning.

D   Cosmids

Cosmids, like other cloning vectors, are plasmids containing antibiotic resistance genes as selective markers and unique restriction sites for cloning. However, in addition to these they contain the DNA sequence from phage λ which is required for packaging into a λ capsid (Collins and Hohn, 1978). This sequence, the cohesive end site or *cos* site, is recognized by the λ packaging system and concatemeric DNA forms produced by ligating linear cosmid and passenger DNA at high concentrations mimic the natural packaging substrate. *In vitro* packaging of the DNA yields a phage preparation which can transduce the cosmid-passenger DNA into λ sensitive cells. Acquisition of the cosmid by the cells can be selected for using the antibiotic resistance markers. The λ head and tail provides a very efficient way of introducing the DNA into bacterial cells; once in the bacterial cytoplasm the cosmids replicate as plasmids and the DNA can be isolated using standard plasmid techniques including chloramphenicol amplification (see reviews of Collins, 1979; Hohn, 1979; Hohn and Hinnen, 1980). *In vitro* packaging and transduction impose a size selection since only those hybrids with a 37—50 kb DNA headful are efficiently recovered (Feiss *et al.*, 1977). This means that only those cosmids which have a large insert of passenger DNA are obtained (Collins and Brüning, 1978; Collins and Hohn, 1978). Currently available cosmids range in size from 6 kb to 24 kb and the optimal DNA insert is one whose size compensates for the difference between the size of the cosmid and a lambda headful (Collins, 1979; Hohn and Collins, 1980).

Cosmids combine the properties of both phage and plasmid vector systems and have advantages over both. The capacity of cosmid vectors to clone large DNA fragments (up to 40 kb for pHC79, see Table 2) is an advantage over the phage vectors. The size selection essentially eliminates the background of cosmids without inserts and greatly reduces the amount of screening needed. In addition, the positive selection for large inserts is in marked contrast to plasmid vectors where the transformation step introduces a bias against the cloning of large fragments (Sherratt, 1979). The major advantage of cosmids is the high efficiency with which input passenger DNA is converted into clones (5×10^4 to 5×10^5 clones per μg passenger DNA). This high cloning efficiency and the capacity to clone large fragments, in conjunction with high density colony hybridization techniques (Hanahan and Meselson, 1980), makes it feasible to construct gene libraries from the entire genome of almost any organism and recover single copy genes. For example, with a library constructed from near-random 40 kb fragments (generated by partial digestion with a frequently cutting restriction enzyme) the number of clones needed to have a 95% probability of finding a particular single copy gene would not be unmanageably large. For *E.coli*, yeast, *Drosophila* and mouse the library would need 1.2×10^2, 4×10^2, 4×10^3 and 7×10^4 cosmid clones respectively.

V   Vectors designed to promote gene expression

This topic is the subject of a separate chapter (Carey, this series, Vol. 4) and will be discussed fairly briefly here. The area has also been reviewed recently by Bernard and Helinski (1980).

A   Gene dosage

Genes from *E.coli* are generally readily expressed when cloned in plasmid vectors in *E.coli* cells. When a high copy number vector is

Table 3 Expression vectors.

Plasmid	Replicon	Size (kb)	Genetic markers	Remarks	Reference
pKN410	R1	15.0	Ap^r	Single *Eco*RI, *Hind*III and *Bam*H1 sites. Runaway replication above 35°C.	1
pKC16	pBR322	11.4	Ap^r	Single *Bam*H1 site. Heat induction of λ replicator increases copy number.	2
pOMPO	pBR322	7.1	Ap^r	*Eco*RI site is downstream of *lac*L8UV5 promoter, following the first 8 amino acids of β-galactosidase.	3, 4
pOP203–13	pBR322	4.9	Ap^r		
pOP95–15	pBR322	4.8	Ap^r	*Eco*RI site downstream of *lac*UV5 promoter between *lac* SD sequence and first amino acid of β-galactosidase.	4
pHUB4	pMK2004	6.5	Km^r	Single *Bam*H1, *Sal*I and *Hpa*I sites downstream of λp_L promoter. Promoter is switched off at 30°C by *cI* repressor. Shift to 42°C inactivates repressor and induces expression of genes linked to p_L.	5
p*trp*ED5–1	pBR322	6.7	Ap^r	Single *Hind*III, *Bam*H1 and *Sal*I sites in *trpD* gene downstream of *trp* promoter. Expression, normally repressed by *trp* repressor, can be induced by 3β-idolylacrylic acid yielding a fusion between part of *trpD* protein and the cloned gene product.	6
pPC01 pPC02 pPC03	pBR322	4.4	Ap^r	*Eco*RI fragments are expressed as fusions with the first 8 amino acids of β-galactosidase in all 3 reading frames. Transcription is from the *lac* promoter.	7

pWT111 pWT121 pWT131	pBR322	4.8	Ap^r	*Hind*III fragments are expressed as fusions with the first 7 amino acids of *trpE* under control of the *trp* promoter.	8
pKT279	pBR322	3.9	Tc^r	One of a series of deletion derivatives of pBR322 with unique *Pst*I sites at various positions in the β-lactamase signal sequence. Available in all 3 reading frames. Fusion protein is transported to the periplasmic space and the signal sequence cleaved off.	9

1, Uhlin *et al.* (1979); 2, Rao and Rogers (1978); 3, Mercereau-Puijalon *et al.* (1978); 4, F. Fuller, manuscript in preparation, 5, Bernard *et al.* (1979); 6, Hallewell and Emtage (1980); 7, Charnay *et al.* (1978); 8, Tacon *et al.* (1980); 9, Talmadge *et al.* (1980a, b).

used, a gene dosage effect is usually observed such that the level of gene product synthesized is increased relative to cells carrying a single copy of the gene on the chromosome (Hershfield *et al.*, 1974; Gelfand *et al.*, 1978). There are, however, some exceptions to this. When the genes for certain ribosomal proteins, or the β and β' subunits of *E.coli* RNA polymerase, are cloned on multicopy plasmids, the increased gene dosage results in a six-fold increase in the level of transcription but the increase in the rate of synthesis of the corresponding proteins is much more modest (one-third to two-fold, Dennis and Fiil, 1979) This is due to a post-transcriptional control system which regulates the cellular level of these proteins (Fiil *et al.*, 1980). Other segments of DNA have proved impossible to clone intact in *E.coli* because they are lethal; the DNA of the early region and replication origin of the lytic coliphage T7 is one such example (Scherzinger *et al.*, 1980).

Aside from these exceptions, gene dosage can be used to increase the yield of protein from a cloned gene. Two systems have been described which allow a conditional increase in vector copy number and hence dosage of the cloned DNA. One uses the "runaway replication" mutant plasmids. These are maintained at 20—50 copies per cell at 30°C (see Table 3, Uhlin *et al.*, 1979) and at temperatures above 35°C replicate without control such that at the end of a 3 h period a 400-fold increase in the plasmid-coded β-lactamase enzyme is observed. In rich medium, heat-induced replication of these "suicide" plasmids is lethal to the cell. The second system uses a derivative of pBR322 which carries a phage λ replicator fragment and the thermolabile *c*I857 repressor gene (see pKC16, Table 3, Rao and Rogers, 1978). A temperature shift induces the λ replication system, causing a three-fold increase in copy number, and since the initial copy number is already high, the over-all effect can be a dramatic increase in yield relative to a single copy chromosomal gene. Rao and Rogers (1978) have observed a 125-fold increase in enzyme yield from a cloned exonuclease III gene. These systems should allow increased yields of many bacterial enzymes.

B   *E.coli* promoters

In contrast to the efficient expression of cloned bacterial genes, those from eukaryotes are expressed poorly if at all in *E.coli*. Genes from yeast and *Neurospora* are the only eukaryotic genes which have been shown to function in *E.coli* without the need of special measures to ensure their expression (Struhl *et al.*, 1976; Ratzkin and Carbon, 1977; Struhl and Davis, 1977; Vapnek *et al.*, 1977; Walz *et al.*, 1978). There may be several reasons for the failure of expression of eukaryotic genes. The introns which separate the coding portions of many eukaryotic genes (Berget *et al.*, 1977; Gilbert, 1978) pose two problems. Firstly the eukaryotic gene may be much larger than the average *E.coli* gene; for example, the mouse dihydrofolate reductase gene spans over 42 kb of DNA (Nunberg *et al.*, 1980). Secondly, *E.coli* has not been shown to be able to splice the introns out of a primary transcript to yield a translatable message. Cloning cDNA copies of mRNAs is one way around this problem, as demonstrated for ovalbumin (Frazer and Bruce, 1978; Mercereau-Puijalon *et al.*, 1978) and other genes. The absence of appropriate nucleotide sequence elements such as promoters and ribosome binding sites may mean that the eukaryotic DNA is not transcribed, or that if it is transcribed, it is not efficiently translated. Finally, if all of the preceding barriers have been successfully avoided, the eukaryotic protein may be degraded by *E.coli* proteases. Expression vectors have been developed which attempt to avoid these barriers in a variety of ways.

Positioning cloned fragments downstream from an efficient *E.coli* promoter has proved an effective way of ensuring transcription. The promoters of the *lac* and *trp* operons, the β-lactamase gene of pBR322, and the leftward promoter of phage λ have been used in this context. The *lac* promoter in particular has been extensively used to drive expression of synthetic somatostatin and human insulin genes (Itakura *et al.*, 1977; Goeddel *et al.*, 1979b), human growth hormone, ovalbumin, β-globin, SV40 t antigen and hepatitis B virus surface antigen (Frazer and Bruce, 1978; Mercereau-Puijalon *et al.*, 1978; Goeddel *et al.*, 1979a; Roberts *et al.*, 1979b; Charney *et al.*, 1980; Guarente *et al.*, 1980). The *lac* promoter is subject to two forms of control, positive regulation by the catabolite gene activator (CAP) system and negative control by the *lac* repressor (Reznikoff and Abelson, 1978). The version of the *lac* promoter most widely used (see Table 3) contains the L8 mutation conferring CAP independence and the UV5 "up-promoter" mutation which enhances the rate of transcription. The high dosage of the *lac* promoter when it is carried on multicopy plasmids overtitrates the *lac* repressor resulting in constitutive *lac* transcription. The promoter can, however, be repressed in strains which over-produce the *lac* repressor (Backman *et al.*, 1976; O'Farrell *et al.*, 1978).

Expression of genes cloned adjacent to the *trp* and λ p_L promoters can be modulated by changing the growth conditions. The *trp* repressor, unlike *lac*, is produced in sufficient quantities to repress even multiple copies of the *trp* promoter. Expression is therefore normally turned off but can be induced by addition of 3 β-indolylacrylic acid (Hallewell and Emtage, 1980). Similarly the λ p_L promoter can be repressed by the *c*I repressor and if a thermolabile repressor is used, then expression can be made dependent upon the temperature of incubation (Bernard *et al.*, 1979). In contrast to the previous three promoters, the activity of the β-lactamase gene promoter cannot be regulated by growth conditions.

C   Fusion proteins

A common strategy to ensure efficient translation of a gene not normally expressed in *E.coli* is to generate a fusion protein by joining the initial portion of a gene which is efficiently translated to the gene whose expression is sought. The ribosome binding site and initiator ATG of the *E. coli* gene then provide the start signals for translation which can proceed into the cloned gene to yield the fusion protein. The correct reading frame has been achieved either by a knowledge of the nucleotide sequence of the cloned fragment and a judicious choice of DNA joining schemes (Itakura *et al.*, 1977; Frazer and Bruce, 1978; Mercereau-Puijalon *et al.*, 1978; Seeburg *et al.*, 1978; Goeddel *et al.*, 1979a, b) or by joining via homopolymeric extensions which naturally generate all three reading frames (Chang *et al.*, 1978; Villa-Komaroff *et al.*, 1978; Burrell *et al.*, 1979; Nagata *et al.*, 1980). More recently, sets of phasing vectors have been developed in which each individual plasmid from a set provides a cloning site in one of the three translational reading frames. Cloning into all three members of a set therefore ensures that one of the derivatives will be framed correctly (see Table 3, Charnay *et al.*, 1978; Tacon *et al.*, 1980; Talmadge *et al.*, 1980b). Phasing vectors have been successfully used to obtain expression of hepatitis B virus surface antigen (Charnay *et al.*, 1980) and fowl plague virus haemagglutinin (Emtage *et al.*, 1980) as fusion proteins with short amino terminal sequences derived from *lac* or *trp* genes. In the case of the thymidine kinase gene of herpes simplex virus, expression was seen regardless of the phase of the cloning vector (Garapin *et al.*, 1980). Sequence analysis of the cloned fragment revealed the presence of fortuitous Shine-Dalgarno sequences (sequences complementary to the 3′ end of the 16 S ribosomal RNA, Shine and Dalgarno, 1975) upstream from the initiator codon for thymidine kinase. A correctly spaced Shine-Dalgarno sequence and initiator codon constitutes a translation start signal (Steitz, 1979) and thus the only requirement for expression of this particular gene is that it is transcribed. Similarly, the mouse dihydrofolate reductase gene is translated from its own initiator codon when cloned by dG:dC tailing into the *Pst*I site of pBR322, even though this site is within the β-lactamase gene. The oligo dG tract preceding the initiation codon may provide some Shine-Dalgarno homology in this case (Chang *et al.*, 1978).

The *Pst*I site in the β-lactamase gene of pBR322 has been widely used for cloning oligo dC tailed DNA fragments since the 3' extension left by *Pst*I cleavage is an ideal substrate for terminal transferase and oligo dG tailing of the vector DNA regenerates *Pst*I sites at the passenger/vector DNA junctions (Chang *et al.*, 1978; Villa-Komaroff *et al.*, 1978; Burrell *et al.*, 1979; Nagata *et al.*, 1980). β-Lactamase is a periplasmic protein with a 23 amino acid signal sequence (Ambler and Scott, 1978; Sutcliffe, 1978) which can serve to transport the bulk of the fusion protein to the periplasmic space (Villa-Komaroff *et al.*, 1978; Talmadge *et al.*, 1980b). Talmadge *et al.* (1980a,b) have shown that a eukaryotic signal sequence, that of preproinsulin, is sufficient to cause transport of the protein to the periplasmic space and the signal sequence is correctly cleaved to yield proinsulin. A periplasmic location may make the product easier to purify and prevent degradation by proteases.

An elegant and versatile method for maximizing expression of a cloned gene has been developed by Ptashne and colleagues (Guarente *et al.*, 1980). It involves positioning a "portable promoter" fragment containing the *lac* promoter and *lac* Shine-Dalgarno sequence at many positions relative to the initiator codon of the cloned gene by *in vitro* nuclease treatments (Backman and Ptashne, 1978; Roberts *et al.*, 1979b). The position which elicits maximal expression is identified in an intermediate step in which the amino-terminal portion of the cloned gene is fused to an enzymatically active carboxy-terminal fragment of β-galactosidase. Expression of the cloned gene can then be easily monitored by measuring β-galactosidase activity. When a high level expression clone is identified the intact gene can be reconstituted by recombination *in vitro* (Guarente *et al.*, 1980). This approach has the advantage that no functional or immunological assays for the cloned product are needed. Using these techniques, expression of SV40 t antigen, human fibroblast interferon and rabbit β-globin as native proteins, unfused to any *E.coli* proteins, has been achieved (Roberts *et al.*, 1979a; Guarente *et al.*, 1980; Taniguchi *et al.*, 1980).

VI   Broad host range vectors

A   *E.coli* and Gram negatives

All of the plasmids discussed so far are restricted to *E.coli* and closely related enteric bacteria such as *Salmonella, Proteus* and *Serratia* spp. The pBR322, pACYC184 and pSC101 replicons, for example, cannot be maintained in *Pseudomonas* strains (Franklin *et al.*, 1981). In contrast, plasmids from the incompatibility groups P and Q can be maintained in almost any Gram negative bacterial species including many strains of potential economic and medical importance such as *Agrobacterium tumefasciens, Pseudomonas, Rhizobium, Azotobacter* and *Neisseria* spp. The development of plasmid vectors for species other than *E.coli* widens the applicability of gene cloning techniques, allowing the planned improvement of strains used in industrial fermentations. An example of this is the enhanced conversion of methanol to single cell protein by *Methylophilus methylotrophus* harbouring a cloned glutamate dehydrogenase gene from *E.coli* (Windass *et al.*, 1980). Vectors for *Pseudomonas* allow the study of the versatile degradative enzymes of these organisms in their normal metabolic background. This is important since some of the *Pseudomonas* enzymes, such as those for toluene degradation, are expressed poorly when cloned in *E.coli* (Jacoby *et al.*, 1978).

The P-group plasmids, many of which were originally isolated from *Pseudomonas* strains, are large plasmids that are capable of conjugative transfer to all of the Gram negative bacterial species that have been tested (Datta and Hedges, 1972; Olsen and Shipley, 1973; Tower and Vivian, 1976). The archetype of the group is RP4, a 56 kb plasmid encoding resistance to ampicillin, kanamycin and tetracycline as well as transfer and replication functions. DNA heteroduplex studies have shown that RP4 is probably identical to the independently isolated plasmids RP1, RK2 and R68 (Burkhardt *et al.*, 1979). Like other broad host range plasmids, RP4 has relatively few sites for restriction enzymes with hexanucleotide recognition sequences considering the size and base composition of its DNA (Meyer *et al.*, 1977; Barth, 1979; Meyer and Shapiro, 1980). Those restriction sites that are present tend to be clustered in the drug resistance genes and this has led to the speculation that intergeneric transfer has selected plasmid molecules best able to overcome restriction barriers, that is, those with the minimal number of restriction targets. The clustering of sites in the drug resistance genes is thought to indicate that these genes have been acquired (by transposition?) relatively recently. A second feature of these plasmids which contrasts with most others is that the genes for replication and maintenance are not tightly clustered but are located in three distinct regions of the genome (Thomas *et al.*, 1980). The genes for conjugal transfer are similarly dispersed (Barth, 1979). The relevance of this to the broad host range is still under investigation but the dispersion of replication functions has complicated the construction of low molecular weight derivatives for use as cloning vectors.

A non-conjugative, non-mobilizable, 11 kb derivative of RK2, pRK2501, has been isolated (see Table 4). Transformation is the

Table 4 Broad host range vectors.

Plasmid	Size (kb)	Genetic markers	Cloning sites/phenotype	Reference
RP4	56.4	Ap^r, Tc^r, Km^r/Tra^+	*Eco*RI, *Bgl*II, *Bam*HI/none *Hind*III/Km^s	1, 2
pRP301	54.7	Ap^r, Tc^r, Tra^+	*Eco*RI, *Bgl*II, *Bam*HI, *Hind*III/none *Sal*I/Tc^s	2
pRK2501	11.1	Tc^r, Km^r	*Eco*RI, *Bgl*II/none *Hind*III, *Xho*I/Km^s *Sal*I/Tc^s	3
pRK290	20.0	Tc^r	*Eco*RI, *Bgl*II/none	4
RSF1010	8.9	Sm^r, Su^r	*Eco*RI, *Hpa*I, *Sst*I/Sm^s *Pvu*II, *Bst*EII/none	5
pKT231	12.3	Km^r, Sm^r	*Hind*III, *Xma*I, *Xho*I/Km^s *Eco*RI, *Sst*I/Sm^s	5
pKT230	11.9	Km^r, Sm^r	as pKT231 plus *Bam*HI/none	5
pKT240	11.6	Km^r, Ap^r	*Hind*III, *Xma*I, *Xho*I/Km^s *Pst*I/Ap^s *Eco*RI, *Hpa*I, *Sst*I, *Bst*EII/none	5
pTB70	17.6	Km^r, Sm^r, Su^r	*Eco*RI, *Sal*I/none	6

Sm^r and Su^r indicate resistance to streptomycin and sulphonamide respectively. Tra^+ indicates ability to promote conjugal DNA transfer.
pRP301, pRK2501 and pRK290 are derived from RP4/RK2. The pKT plasmids are derived from RSF1010 and pTB240 from R300B.
1, Barth and Grinter (1977); 2, Barth (1979); 3, Kahn *et al.* (1979); 4, Bernard and Helinski (1980); 5, Franklin *et al.* (1981); 6, Windass *et al.* (1980).

only way of moving this plasmid between strains and this may pose problems for cloning in species like *Rhizobium* which transform only with difficulty.

Mobilizable vectors circumvent this problem. A second low molecular weight derivative of RK2 is pRK290, which forms the cloning vector half of a binary system for broad host range cloning. The other half of the system is a plasmid that can provide the RK2 transfer gene products in *trans* and thus bring about mobilization of the cloning vector. The second plasmid is a hybrid ColEI replicon carrying the RK2 transfer genes (the genes that were deleted during the size reduction steps used to construct pRK290, Ditta *et al.*, 1980). It is not necessary to introduce this mobilizing plasmid into every clone that is isolated since a triparental mating can be set up comprising two *E.coli* strains, one carrying the mobilizing plasmid and the other the cloning vector hybrid; the third strain in the mixture is the recipient of choice. The mobilizing plasmid will be transferred into the strain containing the cloning vector and cause its mobilization to the final recipient. The narrow host range of the ColEI replicon effectively confines the mobilizing plasmid to the *E.coli* donor strains while allowing efficient intergeneric transfer of the cloning vector component. Because cloning of DNA into the *Eco*RI and *Bgl*II sites available on pRK290 does not give insertional inactivation of any plasmid markers it is recommended that the cleaved vector DNA be phosphatase-treated before ligation to reduce the background of molecules without inserts (Ditta *et al.*, 1980).

The second class of wide host range plasmids are those from incompatibility group Q. They are small, multicopy, mobilizable plasmids of which the archetypes are the closely related plasmids RSF1010 and R300B (Table 4). Derivatives of RSF1010 with alternative drug resistance markers and cloning sites offering insertional inactivation of marker genes have been developed by Timmis (Table 4). They can be readily mobilized by RK2, allowing convenient intergeneric transfer. Efficient transfer is essential if plasmids are to be mobilized into strains which express a DNA restriction system.

B   Bifunctional *Bacillus-Escherichia* vectors

The Gram positive soil organism *Bacillus subtilis* is the most intensively studied bacterium other than *E.coli*. Species of *Bacillus* are widely used in the fermentation industry to produce a variety of exoenzymes (Priest, 1977) and *B. subtilis* would seem to be a good candidate for the production of prokaryotic and eukaryotic proteins from cloned genes. Favourable points are that this organism does not produce disease in uncompromized humans, indeed it is eaten in large quantities in the Orient (Young, 1980), and the secreted product will not be contaminated with endotoxin since this is not a component of the *Bacillus* cell wall as it is in Gram negative cell walls. At present however, the plasmid vectors available for use in *B. subtilis* are at a relatively primitive stage of development compared with *E.coli* vectors. They have been recently reviewed by Lovett and Keggins (1979) and Dubnau *et al.* (1980). Bacilli are spore formers and asporogenous mutant strains have been developed as cloning hosts (Ellis and Dean, 1981). A disabled strain equivalent to *E.coli* χ 1776 has also been constructed (Burke and Le, 1980).

The plasmids found in *B. subtilis* or the closely related *B. pumilis* are almost all either cryptic or carry markers that are not easily selectable. The way towards the development of vector plasmids was opened by the observation that small, high copy number antibiotic resistance plasmids from *Staphylococcus aureus* can be transformed directly into *B. subtilis*. The plasmids are stable and express their drug resistance genes in the new host (Ehrlich, 1977). After reisolation from the initial *Bacillus* strain, these plasmids transform other strains with a frequency of $10^3 - 10^4$ transformants per microgram of DNA; this is rather poor compared with plasmid transformation in *E.coli* and has implications for shotgun cloning. *In vitro* recombination of DNA from a variety of *Staphylococcus* plasmids has yielded a series of plasmids with two or three drug resistance markers, some containing unique cloning sites which give insertional inactivation of the marker (Gryczan and Dubnau, 1978; see list in Dubnau *et al.*, 1980).

Using the 4.5 kb kanamycin resistant plasmid pUB110, Keggins *et al.* (1978) were able to clone *Eco*RI fragments of *B. subtilis* chromosomal DNA which complemented the *trpC2* mutation. In *recE4* host cells the complementing fragment was maintained on the plasmid, but if the cells were rec^+ then the trp^+ marker was recombined into the chromosome. Subsequently, other workers have had great difficulty in shotgun cloning chromosomal fragments. As pointed out by Dubnau *et al.* (1980), this may be due to an inherent property of the transformation of competent *B. subtilis* by plasmids, namely, the requirement for oligomeric plasmid DNA. Canosi *et al.* (1978) showed that it is the oligomeric forms in native plasmid DNA preparations that are active in transformation; supercoiled monomer molecules are inactive. This requirement for oligomers has been explained in terms of present knowledge of DNA uptake by competent bacilli. When duplex DNA is taken up, one strand is degraded and the other enters the cell. A model has been proposed whereby the uptake of oligomeric strands permits intracellular annealing and repair which can regenerate a duplex monomer supercoiled molecule

(Canosi *et al.*, 1981). Any attempt to increase the yield of hybrid molecules in ligation mixtures by increasing the ratio of passenger to vector DNA will decrease the yield of vector oligomers and hence the transformation frequency. This, together with the low over-all transformation frequency is the cause of the difficulties in shotgun cloning. There are several potential ways around this problem.

Chang and Cohen (1979) have reported a highly efficient method of transforming *B. subtilis* protoplasts with plasmid DNA. The procedure involves polyethylene glycol-induced DNA uptake and yields 4×10^7 transformants per microgram of supercoiled DNA. Oligomers are not required; however, there seems to be a preference for lower molecular weight DNA and ligated DNA transforms poorly. Adjusting the DNA concentration during ligation may improve the yield although there may still be problems with homologous fragments since *recE4* protoplasts transform poorly (Dubnau *et al.*, 1980).

A second possible solution would be cloning by recombination. This approach uses a host strain containing a plasmid which shares DNA sequence homology with the vector plasmid used in the ligation reaction. The ligated DNA, even if partially degraded during uptake, can be rescued by recombination with the resident plasmid (Gryczan *et al.*, 1980). There is no requirement for oligomers in this approach, although it is *recE*-dependent. Reported yields are still rather low, being in the range of 6—200 recombinant plasmids per microgram of chromosomal DNA.

In the light of these limitations to using *B. subtilis* as a primary cloning host the possibility of using an efficient primary host such as *E.coli* becomes attractive. It is here that the bifunctional vectors that can replicate in either *B. subtilis* or *E.coli* become most useful. These are double replicon plasmids made *in vitro* by joining plasmids able to replicate in *Bacillus* with *E.coli* vectors such as pBR322. Plasmids from either *Staphylococcus* or *B. cereus* have been used for the *Bacillus* half of the hybrid (Ehrlich, 1978; Kreft *et al.*, 1978). Kreft *et al.* (1978) find that these plasmids are more stable in *E.coli* than in *B. subtilis* hosts and it remains to be seen whether or not this will prove a major problem.

The classification of bacteria on the basis of their reaction to the Gram stain seems to have some relevance for their plasmids since no natural plasmids have been reported which replicate efficiently in both Gram negative and Gram positive hosts. Goze and Ehrlich (1980) have recently shown that some, but not all, *Staphylococcus* plasmids can replicate, albeit rather inefficiently, in *E.coli* cells. They joined a *Staphylococcus* plasmid encoding chloramphenicol resistance to pBR322 and showed that the presence of the *Staphylococcus* replicon allowed the hybrid to replicate under conditions (absence of DNA polymerase I) in which the pBR322 replicon was inactive. This double replicon approach was necessary because the *Staphylococcus* chloramphenicol resistance gene was poorly expressed in *E.coli* so that direct transformation with the *Staphylococcus* plasmid alone and selection for chloramphenicol resistance always failed. The bifunctional vectors are ideal for the study of such heterospecific gene expression.

The number of genes that have been tested for expression in other bacterial species is small. The β-lactamase gene from *S. aureus* can be expressed in *E.coli*, making the cells penicillin resistant (Chang and Cohen, 1974). Staphylococcal chloramphenicol and tetracycline resistance genes are also expressed with varying efficiency (Ehrlich, 1978; Kreft *et al.*, 1978). *B. subtilis* genes for the biosynthesis of leucine and pyrimidines have been shown to function in *E.coli* (Ehrlich *et al.*, 1976; Mahler and Halvorson, 1977; Chi *et al.*, 1978) and Brammar *et al.* (1980) have cloned the β-lactamase from *B. licheniformis* into a phage λ vector and shown a low level of enzyme synthesis in *E.coli*. The level was about 600-fold lower than in the normal host, based on comparison of specific activity; they suggest that it may be inefficient translation of the messenger RNA that is the rate limiting step.

The reciprocal approach has shown that out of the four drug resistance genes tested, only the tetracycline resistance gene of pBR322 is expressed efficiently enough to give a drug resistant phenotype in *B. subtilis* (Kreft *et al.*, 1978, 1981; Ehrlich, 1978). The Tet gene seems to be transcribed by readthrough from a promoter in the *Bacillus* portion of the vector. Northern blots have shown that *E.coli* Ap[r] and Cm[r] genes fail to be transcribed in *B. subtilis* (Kreft *et al.*, 1981). The study of gene expression in different cellular backgrounds is still in its infancy but the bifunctional vectors will certainly have a role to play in further investigations.

E.coli and *B. subtilis* are not the only pair of organisms that have been linked by bifunctional cloning vectors. *E. coli*—yeast vectors are discussed in the chapter by Beggs in Volume 2 of this series. Plasmid rescue experiments are a step towards the development of vectors to maintain genes in either *E.coli* or animal cells. They involve the integration of a hybrid DNA molecule composed of pBR322, a thymidine kinase gene and the early region of SV40, into the genome of a mouse cell line and its subsequent recovery by cloning back into *E.coli* (Hanahan *et al.*, 1980; Perucho *et al.*, 1980). A bifunctional vector for animal cells and *E.coli* would combine the ease of DNA isolation and manipulation in *E.coli* with the ability to reintroduce genes readily into eukaryotic cells to study the effect of *in vitro* modifications.

Before leaving wide host range cloning vectors, mention should be made of the tumour inducing, or Ti, plasmids of *Agrobacterium tumefasciens*. These large conjugative plasmids induce tumours in dicotyledonous plants by transferring part of the plasmid DNA into the plant cell where it becomes stably integrated into the plant chromosomal DNA (Chilton *et al.*, 1977). The possibility of exploiting this DNA transfer system for the genetic engineering of higher plants is discussed in the review by Drummond (1979).

VII Single-stranded DNA phages as cloning vectors

A   Filamentous SS DNA phage biology

The mature virions of the coliphages M13, fd and fl are filaments about a micrometre in length, containing a single stranded circular DNA molecule of 6.4 kb. The DNA strand packaged into the virion is always the plus strand of the double stranded replicative form that is found within infected cells. It is this *in vivo* strand separation and packaging into an easily purified form that is the key feature of the M13 vector systems.

The complete nucleotide sequences of M13 and fd are known; they are 97% homologous. Phage fl is also very closely related to M13 and the three phages can therefore be considered to be equivalent (Beck *et al.*, 1978; Wezenbeek and Schoenmakers, 1979). An outline of the salient points of their infectious cycle will be given here; more detailed information can be found in the monograph entitled "The Single-Stranded DNA Phages" (Denhardt *et al.*, 1978).

M13 absorbs to the sex pili produced by male *E.coli* cells carrying an F plasmid. The requirement for this receptor makes the phage male-specific although infections can be established in female cells if the receptor requirement is circumvented by introducing the phage DNA by transformation. The incoming single-stranded DNA molecule is converted to a duplex replicative form which replicates to give a steady state pool of 50—100 molecules per cell. Progeny single stranded circular molecules accumulate and are packaged and extruded through the cell wall. Release of the progeny phage particles is not lethal to the cell; the main consequences of M13 infection are a 2—3-fold reduction in growth rate and a constant shedding of phage particles into the medium. M13 does not form plaques but zones of slower growing chronically infected cells which look like turbid plaques in a lawn of uninfected cells. The chronically infected cells can be purified by single colony isolation and grown in bulk in liquid culture. The double stranded replicative form can be isolated from the cells by standard plasmid purification techniques and the phage particles can be recovered from the culture supernatant in yields of 5×10^{12} per ml. Because of their filamentous form, the phage particles are easily precipitated by low concentrations of polyethylene glycol; phenol extraction and ethanol precipitation of this material yields DNA of sufficient purity for use as a template in DNA sequencing reactions (Schreier and Cortese, 1979).

There is no defined limit to the size of the DNA molecule that can be packaged into an M13 particle; molecules with inserts of additional DNA simply yield longer phage filaments (Herrmann *et al.*, 1980). In practice however, inserts of above about 5 kb in size tend to be unstable (see below). The size of the phage particles in different isolates can be compared directly by agarose gel electrophoresis and Coomassie blue staining; the visualized phage bands can be recovered from the gel and are still infectious (Herrmann *et al.*, 1978). Alternatively the size of the DNA molecules released by SDS from the phages in as little as $10 \, \mu l$ of culture supernatant can be estimated on an agarose gel (Gardner *et al.*, 1981). The relative orientation of passenger and vector DNA determines which strand of a passenger insert appears in the progeny M13 particles. Clones with the same insert in opposite orientation can be readily identified by mixing two or more phage preparations, releasing the DNA with SDS and allowing time for annealing of homologous DNA strands. DNA from clones with inserts in opposite orientation will be able to form partially double stranded structures since the homologous strands of the passenger DNA will form duplexes (the M13 regions of these molecules will remain single stranded). The partially duplex molecules so formed have reduced agarose gel mobilities compared with non-annealed molecules and this technique provides a relatively simple screen for clones with opposite orientation inserts (Herrmann *et al.*, 1980; Gardner *et al.*, 1981).

B   M13 vectors and their uses

M13 has 10 known genes and the only sizeable intercistronic space is the 508 bp stretch between genes II and IV which contains the origin of DNA replication. It is within this region that most M13 vectors have their cloning sites (see review by Barnes, 1980). The most versatile vectors currently available are listed in Table 5. The RF DNA is isolated and used in ligation reactions exactly as a plasmid and ligated DNA is introduced into a suitable host by transformation of $CaCl_2$ treated cells. Transformants can be isolated either as plaques or, in the case of vectors containing drug resistance markers, as single colonies on appropriate selective plates.

Table 5   M13 vectors.

Phage	Phenotypic marker	Cloning sites/phenotype	Reference
M13mp2	blue plaques	*Eco*RI/white plaques	1
M13mp5	blue plaques	*Hind*III/white plaques	2
M13mp7	blue plaques	*Eco*RI, *Bam*HI, *Pst*I, *Sal*I, *Acc*I, *Hinc*II/white plaques	3
fd101	Ap^r, Km^r	*Pst*I/Ap^s *Hind*III, *Sma*I/Km^s	4
fd103	Ap^r, Cm^r	*Pst*I/Ap^s *Eco*RI/Cm^s	4
fd106	Cm^r, Km^r	*Eco*RI/Cm^s *Xho*I, *Sma*I, *Hind*III/Km^s	4
fd107	Ap^r	*Pst*I/Ap^s *Sal*I, *Hind*III, *Eco*RI/none	4
R199(f 1)	none	*Eco*RI/none	5
M13ho176	*hisD*$^+$	*Pvu*II/*hisD*$^-$ *Eco*RI, *Sal*I, *Pst*I, *Kpn*I/none	6
fdTet	Tc^r	*Eco*RI, *Hind*III, *Xba*I, *Ava*I/none	7

1, Gronenborn and Messing (1978); 2, Messing (1979); 3, Messing *et al.* (1981); 4, Herrmann *et al.* (1980); 5, Boeke *et al.* (1979); 6, Barnes (1979); 7, Zacher *et al.* (1980).

The phages M13mp2, 5 and 7 have a built-in screen to allow visual identification of plaques produced by phages containing passenger DNA. Under appropriate conditions, plaques from empty vectors are blue and those from vectors with inserts are colourless or "white". The vectors contain a fragment of *E.coli* DNA encompassing the *lac* operon regulatory region and the first 145 codons of β-galactosidase which can be translated to yield the α-peptide. The α-peptide, although enzymatically inactive, will complement the *lacZΔM15* deletion which removes the amino terminal region of β-galactosidase. The two special host strains, JM101 and JM103, constructed for use with these vectors carry an F*lac* plasmid containing the *lacZ ΔM*15 deletion and a *traD36* mutation (the latter mutation is a safety feature which allows the cells to make F pili but removes their ability to transfer DNA by conjugation: Messing, 1979; Messing *et al.*, 1981). Infection of these cells leads to α-complementation and the production of active β-galactosidase. On plates containing the *lac* operon inducer IPTG and the chromogenic substrate 5-bromo-4-chloro-indolyl-β-D-galactoside (Xgal) blue plaques will be produced (Gronenborn and Messing, 1978). By an elegant series of DNA manipulations, Messing has introduced an array of cloning sites into codon six of the β-galactosidase α-segment (see Table 5). Cloning into these sites abolishes α-complementation and gives the blue plaque → white plaque phenotypic change.

It should be noted that the earlier of the two hosts for this system, JM101, still has an active K-restriction enzyme and that fragments carrying the unmodified K-recognition sequence AACN$_6$GTGC will be selected against (Kan *et al.*, 1979). The later host, JM103, contains the *hsdR4* mutation which inactivates K-restriction (Messing *et al.*, 1981). Both of these hosts carry an amber suppressor since, as an additional safety feature, the Messing phages have amber mutations in two essential genes.

Preparations of pure single DNA strands, made available in large amounts by M13 cloning, can be used in at least three types of experiment: DNA sequencing, site-directed mutagenesis and nucleic acid hybridization studies.

The rapid and simple method for determining DNA sequences using DNA polymerase and dideoxy chain terminators, as described by Sanger *et al.* (1977), requires a single-stranded template and a suitable primer. Cloning the DNA to be sequenced in M13 provides an ideal template. The primer can either be a denatured restriction fragment from within the cloned segment (Cordell *et al.*, 1979) or, in the case of the Messing vectors, a flanking primer homologous to the vector DNA adjacent to the cloning sites (Heidecker *et al.*, 1980; Messing *et al.*, 1981). Sequence data can be confirmed by sequencing the complementary strand from a clone with the opposite insert orientation. For larger sequencing projects a shotgun sequencing approach has been advocated (Sanger *et al.*, 1980; Messing *et al.*, 1981). This aims to generate banks of clones containing small fragments generated by frequently-cutting restriction enzymes. Clones are selected and sequenced at random and the sequence assembled by finding overlaps using computer analysis to interpret and store the data. Accumulation of data in this way is initially very rapid: 1000 nucleotides per day once templates have been prepared is claimed by Gardner *et al.* (1981), and, initially at least, a restriction map is not required. The endgame, however, is slower as new data accumulate more slowly with increasing amounts of redundant information. A more systematic approach may be needed to complete a sequence: for example, the use of hybridization probes to identify rare clones in the bank or the use of internal primers on larger cloned fragments. The method has been used successfully on the genome of cauliflower mosaic virus (8301 bp, Gardner *et al.*, 1981). Whether larger genomes like phage λ or adenovirus can be totally sequenced by this shotgun approach is not yet known.

Given a knowledge of the wild-type DNA sequence, single stranded DNA from M13 clones can be used to introduce defined nucleotide sequence changes at specific positions by site-directed mutagenesis. This is illustrated by an investigation of the role of the TATA box in the *in vitro* transcription of the chicken conalbumin gene. Wasylyk *et al.* (1980) chemically synthesized an eleven-base oligonucleotide complementary to the DNA sequence across the TATA box in 10 out of 11 positions; the second T of TATA was mispaired with a C residue in the eleven-mer. The eleven-base oligonucleotide, annealed to a wild-type single strand sequence obtained on an M13 vector, then served as a primer for conversion of the molecule to a fully duplex form by DNA polymerase I. Transformation of *E.coli* and replication of the mis-matched molecule generated some progeny in which the TATA sequence had been changed to TAGA (in this particular case these were easy to identify since they generated a new *Xba*I site in the molecule). Mutant clones were found at a frequency of 4% and the T to G change was shown to decrease drastically the *in vitro* transcription of the gene. The increasing availability of synthetic oligonucleotides of defined sequence means that this method will be used more widely.

There have been reports of instability of larger cloned fragments in some M13 vectors and this remains a potential drawback to using these vectors for fragments over 5 kb. Cordell *et al.* (1979) found that 9.4 kb and 5.4 kb fragments of rat DNA spanning the insulin gene were unstable in M13mp5 although a 1.7 kb fragment from the same region proved to be stable. The loss of DNA from fd vectors with large inserts and the accumulation of pseudo-wild-type revertants during propagation has been reported by Herrmann *et al.* (1980). They have been completely unable to clone a 24 kb fragment from plasmid R6-5 or intact phage λ DNA (49 kb) using fd vectors. Finally, Barnes (1980) has found that multiple deletions occurred in a fragment of histidine operon DNA when cloned in M13mp2 at nucleotide 5868 of M13, whereas the same fragment is stable when cloned into a site at nucleotide 5727 to yield the M13Ho176 phage. It should be noted, however, that fragments of less than 3 kb have been cloned and sequenced and the sequence found to be identical to that obtained by the Maxam and Gilbert technique from plasmid clones (Messing *et al.*, 1981). There is, as yet, no evidence of instability of fragments in this size range.

The flexibility of the M13 system for sequencing, and particularly for mutagenesis, makes an M13 vector designed to give expression of a cloned gene desirable. Existing plasmid expression systems could no doubt be adapted for this role. Such a system would allow ready manipulation of the amino acid sequence of the gene product by the directed mutagenesis approach discussed above. Sequence variants could then be used to study the affect of amino acid substitutions on the conformation and biological activity of the protein.

VIII Summary

Plasmid vectors have come a long way since pSC101 in 1973, and the availability of large amounts of cloned DNA has revolutionized many experimental systems. The well studied control elements of *E.coli* operons, like *lac* and *trp*, have been harnessed to accomplish expression of ectopic genes. The construction of libraries of genomic and cDNA clones has been facilitated by the development of cosmid and direct selection vectors. Systems for cloning in hosts other than *E.coli* are being developed and the bifunctional vectors capable of replication in diverse hosts are potentially very powerful. The M13 vectors offer rapid DNA sequencing, targeted mutagenesis and pure preparations of single DNA strands. On all of these fronts we can expect to seen an expansion and refinement of the versatile repertoire of plasmid vectors.

IX Acknowledgements

Thanks are due to Dave Sherratt, Duncan McGeoch and Neil Wilkie for discussions and comments on the manuscript.

X References

Ambler, R. P. and Scott, G. K. (1978). Partial amino acid sequence of penicillinase coded by *Escherichia coli* plasmid R6K. *Proc. Natn. Acad. Sci. U.S.A.* 75, 3732—3736.

An, G. and Friesen, J. D. (1979). Plasmid vehicles for direct cloning of *Escherichia coli* promoters. *J. Bacteriol.* 140, 400—407.

Avni, H. and Markovitz, A. (1979). Characterisation of a mini ColE1 cloning vector. *Plasmid* 2, 225—236.

Backman, K. and Ptashne, M. (1978). Maximising gene expression on a plasmid using recombination *in vitro*. *Cell* 13, 65—71.

Backman, K., Ptashne, M. and Gilbert, W. (1976). Construction of plasmid carrying the cI gene of bacteriophage λ. *Proc. Natn. Acad. Sci. U.S.A.* 73, 4174—4178.

Backman, K., Betlach, M., Boyer, H. W. and Yanofsky, S. (1978). Genetic and physical studies on the replication of ColE1-type plasmids. *Cold Spring Harb. Symp. Quant. Biol.* 43, 69—76.

Bahl, C. P., Marians, K. J., Wu, R., Stawinsky, J. and Narang, S. A. (1976). A general method for inserting specific DNA sequences into cloning vehicles. *Gene* 1, 81—92.

Barnes, W. M. (1979). Construction of an M13 histidine-transducing phage: a single stranded cloning vehicle with one *Eco*RI site. *Gene* 5, 127—139.

Barnes, W. M. (1980). DNA cloning with single-stranded phage vectors. *In* "Genetic Engineering" (Eds. J. K. Setlow and A. Hollaender) Vol. 2, 185—200. Plenum Press, New York.

Barth, P. T. (1979). RP4 and R300B as wide host-range plasmid cloning vehicles. *In* "Plasmids of Medical, Environmental and Commercial Importance" (Eds K. N. Timmis and A. Puhler) 399—410 Elsevier/North-Holland Biomedical Press, Amsterdam.

Barth, P. T. and Grinter, N. J. (1977). Map of plasmid RP4 derived by insertion of transposon C. *J. Molec. Biol.* **113**, 455—474.

Bastia, D. (1978). Determination of restriction sites and the nucleotide sequence surrounding the relaxation site of ColE1. *J. Molec. Biol.* **124**, 601—639.

Beck, E., Sommer, R., Auerswald, E. A., Kurz, C., Zink, B., Osterburg, G. and Schaller, H. (1978). Nucleotide sequence of bacteriophage fd DNA. *Nucl. Acids Res.* **5**, 4495—4503.

Berget, S. M., Moore, C. and Sharp, P. A. (1977). Spliced segments at the 5' terminus of adenovirus 2 late mRNA. *Proc. Natn. Acad. Sci. U.S.A.* **74**, 3171—3175.

Bernard, H. U. and Helinski, D. R. (1980). Bacterial plasmid cloning vehicles. *In* "Genetic Engineering" (Eds J. K. Setlow and A. Hollaender) Vol. 2, 133—167. Plenum Press, New York.

Bernard, H. U., Remaut, E., Hershfield, M. V., Das, H. K. and Helinski, D. R. (1979). Construction of plasmid cloning vehicles that promote gene expression from the lambda P_L promoter. *Gene* **5**, 59—76.

Betlach, M. C., Hershfield, V., Chow, L., Brown, W., Goodman, H. M. and Boyer, H. W. (1976). A restriction endonuclease analysis of the bacterial plasmid controlling the *Eco*RI restriction and modification of DNA. *Fedn Proc. Fedn Am. Socs exp. Biol.* **35**, 2037.

Bird, R. E. and Tomizawa, J. (1978). Ribonucleotide—deoxyribonucleotide linkages at the origin of DNA replication of colicin E1 plasmid. *J. Molec. Biol.* **120**, 137—143.

Bochner, B. R., Huang, H., Schieven, G. L. and Ames, B. N. (1980). Positive selection for loss of tetracycline resistance. *J. Bacteriol.* **143**, 926—933.

Boeke, J. D., Vovis, G. F. and Zinder, N. D. (1979). Insertion mutant of bacteriophage f1 sensitive to *Eco*RI. *Proc. Natn. Acad. Sci. U.S.A.* **76**, 2699—2702.

Bolivar, F. (1978). Construction and characterisation of new cloning vehicles. III. Derivatives of plasmid pBR322 carrying unique *Eco*RI sites for selection of *Eco*RI generated recombinant DNA molecules. *Gene* **4**, 121—136.

Bolivar, F. (1979). Molecular cloning vectors derived from the ColE1-type plasmid pMB1. *Life Sciences* **25**, 807—817.

Bolivar, F. and Backman, K. (1980). Plasmids of *Escherichia coli* as cloning vectors. *In* "Methods in Enzymology" (Ed. R. Wu) Vol. 68, 245—267. Academic Press, New York.

Bolivar, F., Rodriguez, R. L., Betlach, M. C. and Boyer, H. W. (1977a). Construction and characterisation of new cloning vehicles. I. Ampicillin-resistant derivatives of the plasmid pMB9. *Gene* **2**, 75—93.

Bolivar, F., Rodriguez, R. L., Greene, P. J., Betlach, M. C., Heyneker, H. L., Boyer, H. W. (1977b). Construction and characterisation of new cloning vehicles. II. A multipurpose cloning system. *Gene* **2**, 95—113.

Boyer, H. W. and Roulland-Dussoix, D. (1969) A complementation analysis of the restriction and modification of DNA in *Escherichia coli. J. Molec. Biol.* **41**, 459—472.

Brammar, W. J. (1979). Safe and useful vector systems. *Biochem. Soc. Symp.* **44**, 13—27.

Brammar, W. J., Muir, S. and McMorris, A. (1980). Molecular cloning of the gene for the β-lactamase of *Bacillus licheniformis* and its expression in *Escherichia coli. Molec. Gen. Genet.* **178**, 217—224.

Brenner, S. (1979). Genetic Manipulation Advisory Group note 9 and supplements.

Broda, P. (1979). "Plasmids." W. H. Freeman, Oxford and San Francisco.

Bukhari, A. I., Shapiro, J. A. and Adhya, S. L. (Eds) (1977). DNA insertion elements, plasmids and episomes. Cold Spring Harbor Laboratory, New York.

Burke, W. F. and Le, H. T. (1980). Characterisation of a *Bacillus subtilis* strain. *Recombinant DNA Tech. Bull.* **3**, 175—194.

Burkhardt, H., Reiss, G. and Pühler, A. (1979). Relationship of group P1 plasmids revealed by heteroduplex experiments: RP1, RP4, R68 and RK2 are identical. *J. Gen. Microbiol.* **114**, 341—348.

Burrell, C. J., MacKay, P., Greenaway, P. J., Hofschneider, P. H. and Murray, K. (1979). Expression in *Escherichia coli* of hepatitis B virus DNA sequences cloned in plasmid pBR322. *Nature, Lond.* **279**, 43—47.

Cabello, F., Timmis, K. and Cohen, S. N. (1976). Replication control in a composite plasmid constructed by *in vitro* linkage of two distinct replicons. *Nature, Lond.* **259**, 285—290.

Canosi, U., Morelli, G. and Trautner, T. A. (1978). The relationship between molecular structure and transformation efficiency of some *S. aureus* plasmids isolated from *B. subtilis. Molec. Gen. Genet.* **166**, 259—267.

Canosi, U., Iglesias, A. and Trautner, T. A. (1981). Plasmid transformation in *Bacillus subtilis*: Effects of insertion of *Bacillus subtilis* DNA into plasmid pC194. *Molec. Gen. Genet.* **181**, 434—440.

Casadaban, M. J. and Cohen, S. N. (1980). Analysis of gene control signals by DNA fusion and cloning in *Escherichia coli. J. Molec. Biol.* **138**, 179—207.

Chang, A. C. Y. and Cohen, S. N. (1974). Genome construction between bacterial species *in vitro*: Replication and expression of *Staphylococcal* plasmid genes in *E.coli. Proc. Natn. Acad. Sci. U.S.A.* **71**, 1030—1034.

Chang, A. C. Y. and Cohen, S. N. (1978). Construction and characterisation of amplifiable multicopy DNA cloning vehicles derived from the P15A cryptic miniplasmid. *J. Bacteriol* **134**, 1141—1156.

Chang, A. C. Y., Nunberg, J. H., Kaufman, R. J., Erlich, H. A., Schimke, R. T. and Cohen, S. N. (1978). Phenotypic expression in *E.coli* of a DNA sequence coding for mouse dihydrofolate reductase. *Nature, Lond.* **275**, 617—624.

Chang, S. and Cohen, S. N. (1979). High frequency transformation of *Bacillus subtilis* protoplasts by plasmid DNA. *Molec. Gen. Genet.* **168**, 111—115.

Charnay, P., Perricaudet, M., Galibert, F. and Tiollais, P. (1978). Bacteriophage lambda and plasmid vectors allowing fusion of cloned genes in each of the three translational phases. *Nucl. Acids Res.* **5**, 4479—4494.

Charnay, P., Gervais, M., Louise, A., Galibert, F. and Tiollais, P. (1980) Biosynthesis of hepatitis B virus surface antigen in *Escherichia coli. Nature, Lond.* **286**, 893—895.

Chi, N-T. W., Ehrlich, S. D. and Lederberg, J. (1978). Functional expression of two *Bacillus subtilis* chromosomal genes in *Escherichia coli. J. Bacteriol.* **133**, 816—821.

Chilton, M. D., Drummond, M. H., Merrlo, D. J., Sciaky, D., Montoya, A. L., Gordon, M. P. and Nester, E. W. (1977). Stable incorporation of plasmid DNA into higher plant cells: the molecular basis of crown gall tumorigenesis. *Cell* **11**, 263—271.

Chinault, A. C. and Carbon, J. (1979). Overlap hybridisation screening: isolation and characterisation of overlapping DNA fragments surrounding the *leu2* gene on yeast chromosome III. *Gene* 5, 111–126.

Clark, A. J. and Warren, G. J. (1979). Conjugal transmission of plasmids. *A. Rev. Genet.* 13, 99–125.

Clewell, D. B. (1972). Nature of ColE1 plasmid replication in *Escherichia coli* in the presence of chloramphenicol. *J. Bacteriol* 110, 667–676.

Cohen, S. N. and Chang, A. C. Y. (1977). Revised interpretation of the origin of the pSC101 plasmid. *J. Bacteriol.* 132, 734–737.

Cohen, S. N., Chang, A. C. Y., Boyer, H. W. and Helling, R. B. (1973). Construction of biologically functional bacterial plasmids *in vitro*. *Proc. Natn. Acad. Sci. U.S.A.* 70, 3240–3244.

Collins, J. (1977) Gene cloning with small plasmids. *Curr. Topics Microbiol. Immunol.* 78, 121–170.

Collins, J. (1979). *Escherichia coli* plasmids packageable *in vitro* in λ bacteriophage particles. *In* "Methods in Enzymology" (Ed. R. Wu) Vol. 68, 309–326. Academic Press, New York.

Collins, J. and Brüning, H. J. (1978). Plasmids usable as gene cloning vectors in an *in vitro* packaging by coliphage λ: "Cosmids". *Gene* 4, 85–107.

Collins, J. and Hohn, B. (1978). Cosmids: A type of plasmid gene-cloning vector that is packageable *in vitro* in bacteriophage λ heads. *Proc. Natn. Acad. Sci. U.S.A.* 75, 4242–4246.

Cordell, B., Bell, G., Tisher, E., DeNoto, F. M., Ullrich, A., Pictet, R., Rutter, W. J. and Goodman, H. M. (1979). Isolation and characterisation of a cloned rat insulin gene. *Cell* 18, 533–543.

Curtiss, R., Pereirea, D. A., Hsu, J. C., Hull, S. C., Clark, J. E., Maturin, L. F., Goldschmidt, R., Moody, R., Inoue, M. and Alexander, L. (1977). Biological Containment. The subordination of *Escherichia coli* K12. *In* "Recombinant Molecules: Impact on Science and Society" (Eds R. R. Beers and E. G. Bassett). Proceedings of the 10th Miles International Symposium, 45–46.

Datta, N. and Hedges, R. W. (1972). Host ranges of R factors. *J. Gen. Microbiol.* 70, 453–460.

Datta, N., Hedges, R. W., Shaw, E. J., Sykes, R. B. and Richmond, M. H. (1971). Properties of an R factor from *Pseudomonas aeruginosa*. *J. Bacteriol.* 108, 1244–1249.

Denhardt, D. T., Dressler, D. and Ray, D. S. (1978). "The Single-Stranded DNA Phages." Cold Spring Harbor Laboratory, New York.

Dennis, P. P. and Fiil, N. P. (1979). Transcriptional and post-transcriptional control of RNA polymerase and ribosomal protein genes cloned on composite ColE1 plasmids in the bacterium *Escherichia coli*. *J. Biol. Chem.* 254, 7540–7547.

Ditta, G., Stanfield, S., Corbin, D. and Helinski, D. R. (1980). Broad host range cloning system for Gram-negative bacteria: Construction of a gene bank of *Rhizobium meliloti*. *Proc. Natn. Acad. Sci. U.S.A.* 77, 7347–7351.

Donoghue, D. J. and Sharp, P. A. (1978). Replication of Colicin E1 plasmid DNA *in vivo* requires no plasmid-encoded proteins. *J. Bacteriol.* 133, 1287–1294.

Dougan, G. M., Saul, M., Warren, G. and Sherratt, D. J. (1978). A functional map of plasmid ColE1. *Molec. Gen. Genet.* 158, 325–327.

Drummond, M. (1979). Crown Gall Disease. *Nature, Lond.* 281, 343–347.

Dubnau, D., Gryczan, T., Contente, S. and Shivakumar, A. G. (1980). Molecular cloning in *Bacillus subtilis*. *In* "Genetic Engineering (Eds J. K. Setlow and A. Hollaender) Vol. 2, 115—131. Plenum Press, New York.

Ehrlich, S. D. (1977). Replication and expression of plasmids from *Staphylococcus aureus* in *Bacillus subtilis*. *Proc. Natn. Acad. Sci. U.S.A.* **74**, 1680—1682.

Ehrlich, S. D. (1978). DNA cloning in *Bacillus subtilis*. *Proc. Natn. Acad. Sci. U.S.A.* **75**, 1433—1436.

Ehrlich, S. D., Bursztyn-Pettegrew, H., Stroynowski, I. and Lederberg, J. (1976). Expression of the thymidylate synthetase gene of the *Bacillus subtilis* phage Phi-3-T in *Escherichia coli*. *Proc. Natn. Acad. Sci. U.S.A.* **73**, 4145—4149.

Ellis, D. M. and Dean, D. H. (1981). Characterisation of a non-reverting, asporogenous strain of *Bacillus subtilis* 168 for use as an HV1 host. *Recombinant DNA Tech. Bull.* **4**, 1—3.

Emtage, J. S., Tacon, W. C. A., Catlin, G. H., Jenkins, B., Porter, A. G. and Carey, N. H. (1980). Influenza antigenic determinants are expressed from haemagglutinin genes cloned in *Escherichia coli*. *Nature, Lond.* **283**, 171—174.

Falkow, S. (1975). "Infectious multiple drug resistance." Pion Books, London.

Feiss, M., Fisher, R. A., Crayton, M. A. and Egner, C. (1977). Packaging of the bacteriophage λ chromosome: Effect of chromosome length. *Virology* **77**, 281—293.

Fiil, N. P., Friesen, J. D., Downing, W. L. and Dennis, P. P. (1980). Post-transcriptional regulatory mutants in a ribosomal protein — RNA polymerase operon of *E.coli*. *Cell* **19**, 837—844.

Franklin, F. C. H., Bagdasarian, M., Timmis, K. N. (1981) Manipulation of degradative genes of soil bacteria. *In* "Microbial degradation of xenobiotics and recalcitrant compounds" (Eds R. Hütter and T. Leisinger) Academic Press, London and New York.

Frazer, A. C. and Curtiss, R. (1975). Production, properties and utility of bacterial minicells. *Curr. Topics Microbiol. Immunol.* **69**, 1—84.

Frazer, T. H. and Bruce, B. J. (1978). Chicken ovalbumin is synthesised and secreted by *Escherichia coli*. *Proc. Natn. Acad. Sci. U.S.A.* **75**, 5936—5940

Fuchs, L. Y., Covarrubias, L., Escalante, L., Sanchez, S. and Bolivar, F. (1980). Characterisation of a site-specific restriction endonuclease *Sph*I from *Streptomyces phaeochromogenes*. *Gene* **10**, 39—46.

Garapin, A. C., Colbère-Garapin, F., Cohen-Solal, M., Horodniceanu, F. and Kourilsky, P. (1980). Expression of the Herpes simplex virus type 1 thymidine kinase gene in *Escherichia coli*. *Proc. Natn. Acad. Sci. U.S.A.* **78**, 815—819.

Gardner, R. C., Howarth, A. J., Hahn, P., Brown-Luedi, M., Shepherd, R. J. and Messing, J. (1981). The complete nucleotide sequence of an infectious clone of cauliflower mosaic virus by M13mp7 shotgun sequencing. *Nucl. Acids Res.* **9**, 2871—2888.

Gelfand, D. H., Shepard, H. M., O'Farrell, P. H., Polisky, B. (1978) Isolation and characterisation of a ColE1-derived plasmid copy-number mutant. *Proc. Natn. Acad. Sci. U.S.A.* **75**, 5869—5873.

Gilbert, W. (1978). Why genes in pieces? *Nature, Lond.* **271**, 501.

Goebel, W. and Schrempf, H. (1972). Isolation of minicircular deoxyribonucleic acid from wild strains of *Escherichia coli* and their relationship to other bacterial plasmids. *J. Bacteriol.* **111**, 696—704.

Goeddel, D. V., Heyneker, H. L., Hozumi, T., Arentzen, R., Itakura, K., Yansura, D. G., Ross, M. J., Miozzari, G., Crea, R. and Seeberg, P. H. (1979a). Direct expression in *Escherichia coli* of a DNA sequence coding for human growth hormone. *Nature, Lond.* **281**, 544—548.

Goeddel, D. V., Kleid, D. G., Bolivar, F., Heyneker, H. L., Yansura, D. G., Crea, R., Hirose, T., Krzenwski, A., Itakura, K. and Riggs, A. D. (1979b) Expression *Escherichia coli* of chemically synthesised genes for human insulin *Proc. Natn. Acad. Sci. U.S.A.* **76**, 106—110.

Goze, A. and Ehrlich, S. D. (1980). Replication of plasmids from *Staphylococcus aureus* in *Escherichia coli. Proc. Natl. Acad. Sci. U.S.A.* **77**, 7333—7337.

Gronenborn, B. and Messing, J. (1978). Methylation of single-stranded DNA *in vitro* introduces new restriction endonuclease cleavage sites. *Nature, Lond.* **272**, 375—376.

Gryczan, T. J. and Dubnau, D. (1978). Construction and properties of chimeric plasmids in *Bacillus subtilis. Proc. Natn. Acad. Sci. U.S.A.* **75**, 1428—1432.

Gryczan, T., Contente, S. and Dubnau, D. (1980). Molecular cloning of heterologous chromosomal DNA by recombination between a plasmid vector and a homologous resident plasmid in *Bacillus subtilis. Molec. Gen. Genet.* **177**, 459—467.

Guarente, L., Lauer, G., Roberts, T. M. and Ptashne, M. (1980). Improved methods for maximising expression of a cloned gene: a bacterium that synthesises rabbit β-globin. *Cell* **20**, 543—553.

Gustafsson, P. and Nordström, K. (1978). Temperature-dependent and amber copy mutants of plasmid R1—19 in *Escherichia coli. Plasmid* **1**, 136—144.

Hallewell, R. A. and Emtage, S. (1980). Plasmid vectors containing the tryptophan operon promoter suitable for efficient regulated expression of foreign genes. *Gene* **9**, 27—47.

Hanahan, D. and Meselson, M. (1980). Plasmid screening at high colony density. *Gene* **10**, 63—67.

Hanahan, D., Lane, D., Lipsich, L., Wigler, M. and Botchan, M. (1980). Characteristics of an SV40-plasmid recombinant and its movement into and out of the genome of a murine cell. *Cell* **21**, 127—139.

Hashimoto-Gotoh, T., Nordheim, A. and Timmis, K. N. (1980). Specific purpose plasmid cloning vectors. I. A low copy number pSC101-derived biological containment vector. Submitted for publication.

Heffron, F., Sublett, R., Hedges, R. W., Jacob, A. and Falkow, S. (1975). Origin of the TEM beta-lactamase gene found on plasmids. *J. Bacteriol.* **122**, 250—256.

Heidecker, G., Messing, J. and Gronenborn, B. (1980). A versatile primer for DNA sequencing in the M13mp2 cloning system. *Gene* **10**, 69—73.

Herrmann, R., Neugebauer, K., Zentgraf, H. and Schaller, H. (1978). Transposition of a DNA sequence determining kanamycin resistance into the single stranded genome of bacteriophage fd. *Molec. Gen. Genet.* **159**, 171—178.

Herrmann, R., Neugebauer, K., Pirkl, E. Zentgraf, H. and Schaller, H. (1980). Conversion of bacteriophage fd into an efficient single-stranded DNA vector system. *Molec. Gen. Genet.* **177**, 231—242.

Hershfield, V., Boyer, H. W., Yanofsky, C., Lovett, M. A. and Helinski, D. R. (1974). Plasmid ColEI as a molecular vehicle for cloning and amplification of DNA. *Proc. Natn. Acad. Sci. U.S.A.* **71**, 3455—3459.

Hohn, B. (1979). *In vitro* packaging of λ and cosmid DNA. *In* "Methods in Enzymology" (Ed. R. Wu) Vol. 68, 299—309. Academic Press, New York.

Hohn, B. and Collins, J. (1980). A small cosmid for efficient cloning of large DNA fragments. *Gene* 11, 291—298.

Hohn, B. and Hinnen, A. (1980). Cloning with cosmids in *E.coli* and yeast. *In* "Genetic Engineering" (Eds J. K. Setlow and A. Hollaender) Vol. 2, 169—183. Plenum Press, New York.

Inselberg, J. and Ware, P. (1979). A complementation analysis of mobilization deficient mutants of the plasmid ColE1. *Molec. Gen. Genet.* 172, 211—219.

Itakura, K., Hirose, T., Crea, R., Riggs, A. D., Heyneker, H. L., Bolivar, F. and Boyer, H. W. (1977). Expression in *Escherichia coli* of a chemically synthesised gene for the hormone somatostatin. *Science, N. Y.* 198, 1056—1063.

Jackson, D. A., Symons, R. H. and Berg, P. (1972). Biochemical method for inserting new genetic information into DNA of Simian virus 40: circular SV40 DNA molecules containing lambda phage genes and the galactosidase operon of *Escherichia coli. Proc. Natn. Acad. Sci. U.S.A.* 69, 2904—2909.

Jacoby, G. A., Rogers, J. E., Jacob, A. E. and Hedges, R. W. (1978). Transposition of *Pseudomonas* toluene-degrading genes and expression in *Escherichia coli. Nature, Lond.* 274, 179—180.

Johnson, D. A. and Willetts, N. S. (1980). F-derived *tra*$^+$ recombinants transfer and transposition properties. *In* "Plasmids: Environmental Effects and Maintenance Mechanisms" (Eds C. Stuttard and S. Light) 293—301. Academic Press, New York.

Jones, I. M., Primrose, S. B., Robinson, A. and Ellwood, D. C. (1980). Maintenance of some ColE1-type plasmids in chemostat culture. *Molec. Gen. Genet.* 180, 579—584.

Kahn, M. and Helinski, D. R. (1978). Construction of a novel plasmid-phage hybrid: use of the hybrid to demonstrate ColE1 DNA replication *in vivo* in the absence of a ColE1 — specified protein. *Proc. Natn. Acad. Sci. U.S.A.* 75, 2200—2204.

Kahn, M., Kolter, R., Thomas, C., Figurski, D., Meyer, R., Remaut, E. and Helinski, D. R. (1979). Plasmid cloning vehicles derived from plasmids ColE1, F, R6K and RK2. *In* "Methods in Enzymology" (Ed. R. Wu) Vol. 68, 268—280. Academic Press, New York.

Kan, N. C., Lautenburger, J. A., Edgell, M. H. and Hutchinson, C. A. (1979). The nucleotide sequence recognised by the *Escherichia coli* K12 restriction and modification enzymes. *J. Molec. Biol.* 130, 191—209.

Keggins, K. M., Lovett, P. S. and Duvall, E. J. (1978). Molecular cloning of genetically active fragments of *Bacillus* DNA in *Bacillus subtilis* and properties of the vector plasmid pUB110. *Proc. Natn. Acad. Sci. U.S.A.* 75, 1423—1427.

Kolter, R., Inuzuka, M. and Helinski, D. R. (1978). Trans-complementation dependent replication of a low molecular weight origin fragment from plasmid R6K. *Cell* 15, 1199—1208.

Kolter, R. and Helinski, D. R. (1979). Regulation of initiation of DNA replication. *A. Rev. Genet.* 13, 355—391.

Kreft, J., Bernard, K. and Goebel, W. (1978). Recombinant plasmids capable of replication in *B. subtilis* and *E.coli. Molec. Gen. Genet.* 162, 59—67.

Kreft, J., Burger, K. J. and Goebel, W. (1981) Studies on the expression of antibiotic resistance genes from *E. coli* in *B. subtilis. Molec. Gen. Genet.* in press.

Lovett, P. S. and Keggins, K. M. (1979) *Bacillus subtilis* as a host for molecular cloning. *In* "Methods in Enzymology" (Ed. R. Wu) Vol. 68, 342—350. Academic Press New York.

Mahler, J., Halvorson, H. O. (1977) Tranformation of *Escherichia coli* and *Bacillus subtilis* with a hybrid plasmid molecule. *J. Bacteriol.* 131, 374—377.

Maloy, S. R., Nunn, W. D. (1981). Selection for loss of tetracycline resistance by *Escherichia coli. J. Bacteriol.* 145, 1110—1112.

Maniatis, T., Hardison, R. C., Lacy, E., Lauer, J., O'Connel, C., Quon, D., Sim, G. K. and Efstratiadis, A. (1978). The isolation of structural genes from libraries of eukaryotic DNA. *Cell* 15, 687—701.

Manis, J. J. and Kline, B. C. (1977). Restriction endonuclease mapping and mutagenesis of the F sex factor replication region. *Molec. Gen. Genet.* 152, 175—182.

Meacock, P. A. and Cohen, S. N. (1980). Partitioning of bacterial plasmids during cell division: a cis-acting locus that accomplishes stable plasmid inheritance. *Cell* 20, 529—542.

Mercerau-Puijalon, O., Royal, A., Cami, B., Garapin, A., Krust, A., Gannon, F. and Kourilsky, P. (1978). Synthesis of an ovalbumin-like protein by *Escherichia coli* K12 harbouring a recombinant plasmid. *Nature, Lond.* 275, 505—510.

Messing, J. (1979). A multipurpose cloning system based on the single-stranded DNA bacteriophage M13. *Recombinant DNA Tech. Bull.* 2, 43—48.

Messing, J., Crea, R. and Seeburg, P. H. (1981). A system for shotgun sequencing. *Nucl. Acids Res.* 9, 309—321.

Meyer, R., Figurski, D. and Helinski, D. R. (1977). Physical and genetic studies with restriction endonucleases on the broad host-range plasmid RK2. *Molec. Gen . Genet.* 152, 129—135.

Meyer, R. J. and Shapiro, J. A. (1980). Genetic organisation of the broad host-range IncP-1 plasmid R751. *J. Bacteriol.* 143, 1362—1373.

Morrow, J. F. (1979). Recombinant DNA techniques. *In* "Methods in Enzymology" (Ed. R. Wu) Vol. 68, 3—24. Academic Press, New York.

Nagata, S., Taira, H., Hall, A., Johnsrud, L., Streuli, M., Escödi, J., Boll, W., Cantell, K. and Weissmann, C. (1980). Synthesis in *E.coli* of a polypeptide with a human leukocyte interferon activity. *Nature, Lond.* 284, 316—320.

Neve, R. L., West, R. W. and Rodriquez, R. L. (1979). Eukaryotic DNA fragments which act as promoters for a plasmid gene. *Nature, Lond.* 277, 324—325.

Norgard, M. V., Emigholz, K. and Monahan, J. J. (1979). Increased amplification of pBR322 plasmid deoxyribonucleic acid in *Escherichia coli* K12 strains RR1 and χ1776 grown in the presence of high concentrations of nucleoside. *J. Bacteriol.* 138, 270—272.

Novick, R. P., Clowes, R. C., Cohen, S. N., Curtiss, R., Datta, N. and Falkow, S. (1976). Uniform nomenclature for bacterial plasmids: a proposal. *Bact. Rev.* 40, 168—189.

Nunberg, J. H., Kaufman, R. J., Chang, A. C. Y. C., Cohen, S. N. and Schimke, R. T. (1980). Structure and genomic organisation of the mouse dihydrofolate reductase gene. *Cell* 19, 355—364.

O'Farrel, P. H., Polisky, B. and Gelfand, D. H. (1978). Regulated expression by readthrough translation from a plasmid-encoded β-galactosidase. *J. Bacteriol.* 134, 645—654.

Oka, A., Nomura, N., Morita, M., Sugisaki, H., Sugimoto, K. and Takanami, M. (1979). Nucleotide sequence of small ColEI derivatives: structure of the regions essential for autonomous replication and colicin E1 immunity. *Molec. Gen. Genet.* 172, 151—159.

Olsen, R. H. and Shipley, P. (1973). Host ranges and properties of the *Pseudomonas aeruginosa* R factor R1822. *J. Bacteriol.* 113, 772—780.

Ozaki, L. S., Maeda, S., Shimada, K. and Takagi, Y. (1980). A novel ColE1::Tn3 plasmid vector that allows direct selection of hybrid clones in *E. coli. Gene* 8, 301—314.

Perucho, M., Hanahan, D., Lipsich, L. and Wigler, M. (1980). Isolation of the chicken thymidine kinase gene by plasmid rescue. *Nature, Lond.* 285, 207—210.

Polisky, B., Greene, P., Garfin, D. E., McCarthy, B. J., Goodman, H. M. and Boyer, H. W. (1975). Specificity of substrate recognition by the *Eco*RI restriction endonuclease. *Proc. Natn. Acad. Sci. U.S.A.* 72, 3310—3314.

Priest, F. G. (1977). Extracellular enzyme synthesis in the genus *Bacillus. Bact. Rev.* 41, 711—753.

Pritchard, R. H., Barth, P. T. and Collins, J. (1969). Control of DNA synthesis in bacteria. *Symp. Soc. Gen. Microbiol.* 19, 263—297.

Rao, R. N. and Rogers, S. G. (1978). A thermoinducible λ phage-ColE1 plasmid chimera for the overproduction of gene products from cloned segments. *Gene* 3, 247—263.

Rao, R. N. and Rogers, S. G. (1979). Plasmid pKC7: A vector containing ten restriction endonuclease sites suitable for cloning DNA segments. *Gene* 7, 79—82.

Ratzkin, B. and Carbon, J. (1977). Functional expression of cloned yeast DNA in *Escherichia coli. Proc. Natn. Acad. Sci. U.S.A.* 74, 487—491.

Reznikoff, W. S. and Abelson, J. N. (1978). The *lac* promoter. *In* "The Operon" (Eds J. H. Miller and W. S. Reznikoff) 221—243. Cold Spring Harbor Laboratory, New York.

Roberts, R. J. (1978). Restriction and modification enzymes and their recognition sequences. *Gene* 4, 183—193.

Roberts, T. M., Bikel, I., Yocum, R. R., Livingston, D. M. and Ptashne, M. (1979a). Synthesis of simian virus 40 t antigen in *Escherichia coli. Proc. Natn. Acad. Sci. U.S.A.* 76, 5596—5600.

Roberts, T. M., Kacich, R. and Ptashne, M. (1979b). A general method for maximising the expression of a cloned gene. *Proc. Natn. Acad. Sci. U.S.A.* 76, 760—764.

Roberts, T. M., Swanberg, S. L., Poteete, A., Riedel, G. and Backman, K. (1980). A plasmid cloning vehicle allowing a positive selection for inserted fragments. *Gene.* 12, 123—127.

Rodriguez, R. L., Bolivar, F., Goodman, H. M., Boyer, H. W. and Betlach M. (1976). Construction and characterisation of cloning vehicle *In* "Molecular Mechanisms in the control of gene expression" (Eds D. P. Nierlich, W. J. Rutler and C. F. Fox) 471—477. Academic Press, New York.

Rodriguez, R. L., West, R. W., Heyneker, H. L., Bolivar, F. and Boyer, H. W. (1979). Characterising wild-type and mutant promoters of the tetracycline resistance gene in pBR313. *Nucl. Acids Res.* 6, 3267—3287.

Rüther, U. (1980). Construction and properties of a new cloning vehicle allowing direct screening for recombinant plasmids. *Molec. Gen. Genet.* 178, 475—477.

Sancar, A., Hack, A. M. and Rupp, W. D. (1979). Simple method for identification of plasmid-coded proteins. *J. Bacteriol.* 137, 692—693.

Sanger, F., Nicklen, S. and Coulson, A. R. (1977). DNA sequencing with chain termination inhibitors. *Proc. Natn. Acad. Sci. U.S.A.* 74, 5463—5467.

Sanger, F., Coulson, A. R., Barrel, B. G., Smith, A. J. H. and Roe, B. A. (1980).

Cloning in single-stranded bacteriophage as an aid to rapid DNA sequencing. *J. Molec. Biol.* **143**, 161—178.

Scherzinger, E., Lauppe, H-F., Voll, N. and Wanke, M. (1980). Recombinant plasmids carrying promoters, genes and the origin of DNA replication of the early region of bacteriophage T7. *Nucl. Acids Res.* **8**, 1287—1305.

Schreier, P. H. and Cortese, R. (1979). A fast and simple method for sequencing DNA cloned in the single-stranded bacteriophage M13 *J. Molec. Biol.* **129**, 169—172.

Schumann, W. (1979). Construction of an *HpaI* and *HindII* plasmid vector allowing direct selection of transformants harbouring recombinant plasmids. *Molec. Gen. Genet.* **174**, 221—224.

Seeburg, P. H., Shine, J., Martial, J. A., Ivarie, R. D., Morris, J. A., Ullrich, A., Baxter, J. D. and Goodman, H. M. (1978). Synthesis of growth hormone by bacteria. *Nature, Lond.* **276**, 795—798.

Shepard, H. M., Gelfand, D. H. and Polisky, B. (1979). Analysis of a recessive plasmid copy number mutant: Evidence for negative control of ColE1 replication. *Cell* **18**, 267—275.

Sherratt, D. J. (1979). Plasmid vectors for genetic manipulation *in vitro*. *Biochem. Soc. Symp.* **44**, 29—36.

Shine, J. and Dalgarno, L. (1975). Determinant of cistron specificity in bacterial ribosomes. *Nature, Lond.* **254**, 34—38.

Soberon, X., Covarrubias, L. and Bolivar, F. (1980). Construction and characterisation of new cloning vehicles. IV. Deletion derivatives of pBR322 and pBR325. *Gene* **9**, 287—305.

Staudenbauer, W. L. (1978). Structure and replication of the colicin E1 plasmid. *Curr. Topics Microbiol. Immunol.* **83**, 93—156.

Steitz, J. A. (1979). Genetic signals and nucleotide sequences in messenger RNA. *In* "Biological Regulation and Development" (Ed. R. F. Goldberger) Vol. 1, 349—389. Plenum Press. New York.

Struhl, K. and Davis, R. W. (1977). Production of a functional eukaryotic enzyme in *Escherichia coli*: Cloning and expression of the yeast structural gene for imidazole—glycerol phosphate dehydratase (*his3*). *Proc. Natn. Acad. Sci. U.S.A.* **74**, 5255—5259.

Struhl, K., Cameron, J. R. and Davis, R. W. (1976). Functional genetic expression of eukaryotic DNA in *Escherichia coli*. *Proc. Natn. Acad. Sci. U.S.A.* **73**, 1471—1475.

Sutcliffe, J. G. (1978a). Complete nucleotide sequence of the *Escherichia coli* plasmid pBR322. *Cold Spring Harb. Symp. Quant. Biol.* **43**, 77—90.

Sutcliffe, J. G. (1978b). pBR322 restriction map derived from the DNA sequence: accurate DNA size markers up to 4361 nucleotide pairs long. *Nucl. Acids Res.* **5**, 2721—2728.

Tacon, W., Carey, N. and Emtage, S. (1980). The construction and characterisation of plasmid vectors suitable for the expression of all DNA phases under the control of the *E. coli* tryptophan promoter. *Molec. Gen. Genet.* **177**, 427—438.

Tait, R. C. and Boyer, H. W. (1978a). Restriction endonuclease mapping of pSC101 and pMB9. *Molec. Gen. Genet.* **164**, 285—288.

Tait, R. C. and Boyer, H. W. (1978b). On the nature of tetracycline resitance controlled by the plasmid pSC101. *Cell* **13**, 78—81.

Tait, R. C., Rodriguez, R. L. and Boyer, H. W. (1977). Altered tetracycline resistance in pSC101 recombinant plasmids. *Molec. Gen. Genet.* **151**, 327—331.

Talmadge, K., Kaufman, J., Gilbert, W. (1980a). Bacteria mature preproinsulin to proinsulin. *Proc. Natn. Acad. Sci. U.S.A.* **77**, 3988—3992.

Talmadge, K., Stahl, S., Gilbert, W. (1980b) Eukaryotic signal sequence transports insulin antigen in *Escherichia coli. Proc. Natn. Acad. Sci. U.S.A.* **77**, 3369—3373.

Taniguchi, T., Guarente, L., Roberts, T. M., Kimelman, D., Douhan, J. and Ptashne, M. (1980). Expression of the human fibroblast interferon gene in *Escherichia coli. Proc. Natn. Acad. Sci. U.S.A.* **77**, 5230—5233.

Thomas, C. M., Meyer, R. and Helinski, D. R. (1980). Regions of broad-host range plasmid RK2 which are essential for replication and maintenance. *J. Bacteriol.* **141**, 213—222.

Timmis, K. N. (1980). *In vitro* methods for the manipulation of DNA. *In* "Genetics as a Tool in Microbiology" (Eds D. A. Hopwood and S. W. Glover). Cambridge University Press.

Towner, K. J. and Vivian, A. (1976). RP4-mediated conjugation in *Acinetobacter calcoaceticus. J. Gen. Microbiol.* **93**, 355—360.

Twigg, A. J. and Sherratt, D. J. (1980). Trans-complementable copy-number mutants of plasmid ColE1. *Nature, Lond.* **283**, 216—218.

Uhlin, B. E., Molin, S., Gustafsson, P. and Nordström K. (1979). Plasmids with temperature-dependent copy number for amplification of cloned genes and their products. *Gene* **6**, 91—106.

Ullrich, A., Shine, J., Chirgwin, J., Pictet, R., Tischer, E., Rutter, W. J. and Goodman, H. M. (1977). Rat insulin genes: construction of plasmids containing the coding sequences. *Science, N.Y.* **196**, 1313—1319.

Vapnek, D., Hautala, J. A., Jacobson, J. W., Giles, N. H. and Kushner, S. R. (1977). Expression in *Escherichia coli* K-12 of the structural gene for catabolic dehydroquinase of *Neurospora crassa. Proc. Natn. Acad. Sci. U.S.A.* **74**, 3508—3512.

Villa-Komaroff, L., Efstratiadis, A., Broome, S., Lomedico, P., Tyard, R., Naber, S. P., Chick, W. L. and Gilbert, W. (1978). A bacterial clone synthesising proinsulin. *Proc. Natn. Acad. Sci. U.S.A.* **75**, 3727—3731.

Walz, A., Ratzkin, B. and Carbon, J. (1978). Control of expression of a cloned yeast (*Saccharomyces cerevisiae*) gene (*trp5*) by a bacterial insertion element (IS2). *Proc. Natn. Acad. Sci. U.S.A.* **75**, 6172—6176.

Warren, G. J., Twigg, A. J. and Sherratt, D. J. (1978). ColE1 plasmid mobility and relaxation complex. *Nature, Lond.* **274**, 259—261.

Wasylyk, B., Derbyshire, R., Guy, A., Molko, D., Roget, A., Teoule, R. and Chambon, P. (1980). Specific *in vitro* transcription of conalbumin gene is drastically reduced by single point mutation in T-A-T-A box homolgy sequence. *Proc. Natn. Acad. Sci. U.S.A.* **77**, 7024—7028.

West, R. W. and Rodriguez, R. L. (1980). Construction and characterisation of *E. coli* promoter—probe plasmid vectors II. RNA polymerase binding studies on antibiotic-resistance promoters. *Gene* **9**, 175—193.

West, R. W., Neve, R. L. and Rodriguez, R. L. (1979). Construction and characterisation of *E. coli* promoter—probe plasmid vectors. I. Cloning of promoter containing DNA fragments. *Gene* **7**, 271—288.

Wezenbeek, P. and Schoenmakers, J. G. G. (1979). Nucleotide sequence of genes III, VI and I of bacteriophage M13. *Nucl. Acids Res.* **6**, 2799—2818.

Widera, G., Gautier, F., Lindenmainer, W. and Collins, J. (1978). The expression of tetracycline resistance after insertion of foreign DNA fragments between the *EcoRI* and *HindIII* sites of the plasmid cloning vector pBR322. *Molec. Gen. Genet.* **163**, 301—305.

Willetts, N. and Skurray, R. (1980). The conjugation system of F-like plasmids. *A. Rev. Genet.* **14**, 41—76.

Windass, J. D., Worsey, M. J., Pioli, E. M., Pioli, D., Barth, P. T., Atherton, K. T., Dart, E. C., Byrom, D., Powell, K. and Senior, P. J. (1980). Improved conversion of methanol to single-cell protein by *Methylophilus methylotrophus*. *Nature, Lond.* **287**, 396—401.

Young, F. E. (1980). Impact of cloning in *Bacillus subtilis* on fundamental and industrial microbiology. *J. Gen. Microbiol.* **119**, 1—15.

Zacher, A. N., Stock, C. A., Golden, J. W. and Smith, G. P. (1980). A new filamentous phage cloning vector: fd-tet. *Gene* **9**, 127—140.

Vectors based on bacteriophage lambda

W. J. BRAMMAR

Department of Biochemistry, University of Leicester, Leicester, UK

I   Introduction

Coliphage lambda has long been a favoured subject for geneticists and a wealth of detailed information about the organization and

function of the lambda genome is now available. The phage's dual
life style, involving either a lytic phase or a stable relationship with
the host in the quiescent, lysogenic phase, has allowed the isolation
of regulatory mutants and strains defective in essential phage functions. The specialized lambdoid transducing phages carrying *gal*, *bio*,
trp or *lac* genes of the host, originally isolated by aberrant excision
of phage genomes from the prophage state, have been key tools in
developing our understanding of molecular genetics. They have also
provided a plethora of technical information on the behaviour of
genes cloned within the lambda genome.

With the advent of restriction enzyme technology in the early
1970s lambda geneticists naturally adapted the phage to become a
useful cloning vector and "transducing phages" could be produced
containing DNA from any source. The lambdoid vectors were slow to
be adopted at first, largely because the baroque nature of lambda
genetics discouraged the non-specialist by its arcane sophistication.
It is precisely this wealth of esoteric detail, appropriately harnessed,
that makes lambda so useful and versatile as a cloning vehicle. This
chapter is intended as a simplifying guide to the use of lambdoid
cloning vectors.

II   The lambda genome

A   The arrangement of the genes

The genome of bacteriophage lambda contains about 50 genes,
though only 50% of these are essential for phage growth and plaque
formation. The genome is packaged into the head of the mature
phage particle as a non-permuted, linear, duplex DNA molecule with
single stranded 5' projections of 12 nucleotides. These mutally
complementary sequences anneal to allow circularization of the
genome immediately on infection.

Genes of related function tend to be clustered on the genome. The
essential genes concerned with head $(A-F)$ and tail $(Z-J)$ formation
and assembly occupy the left-hand third of the genome (Fig. 1). The
region of the map between gene J and *att*, the phage attachment site,
contains non-essential genes coding for proteins of unknown or
unimportant function (Hendrix, 1971). Genes to the right of *att*
govern site-specific (*int* and *xis*) and generalized (*red A* and *red B*)
recombination of phage DNA and the control of lysogeny (*cIII*).
None of the genes between J and N is essential for the phage's ability
to grow lytically and make plaques, and it is largely this region of
the genome that is deleted or replaced when using lambda as a
vector. The product of gene N is the early regulatory protein that is

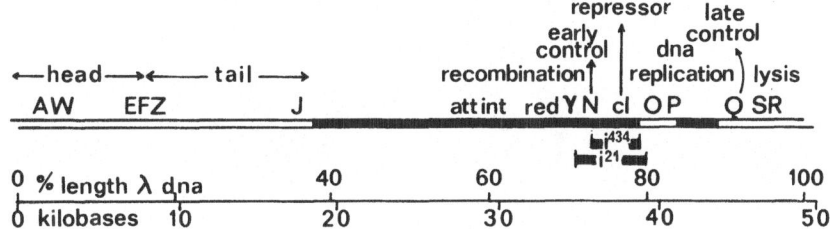

Figure 1 Simplified map of the bacteriophage lambda genome. Genes in the unshaded regions are essential for phage growth and plaque formation: those in the shaded areas are non-essential. Phage genes mentioned in the text are indicated, as are the positions of the frequently used immunity-substitutions, i^{434} ($= imm^{434}$) and i^{21} ($= imm^{21}$)

normally necessary to activate transcription of most other phage genes.

Gene *cI* is the structural gene for the lambda phage repressor, the protein that switches off transcription of phage genes in the lysogenic state (Ptashne, 1971). The presence of the λ repressor makes a λ-lysogenic cell immune to superinfection by another λ phage and is responsible for the characteristic turbidity of lambda plaques. The genes to the right of *cI* in Fig. 1 are all essential for plaque formation. The products of genes *O* and *P* are required for replication of λ DNA (Brooks, 1965; Joyner *et al.*, 1976), the *Q* gene product for activation of late transcription (Dove, 1966; Couturier *et al.*, 1973) and the *S* and *R* gene products for lysis of the host cells (Campbell and Del Campillo-Campbell, 1963; Harris *et al.*, 1967).

B   The expression of lambda genes

The lambda genome contains only three promoters that are important for expression of genes during the lytic cycle. Several other promoters are involved in events during the lysogenic response, but these are not germane to our present purposes. Immediately after infection of a sensitive host or induction of a lysogenic strain, transcription by the host's RNA polymerase proceeds leftwards from promoter p_L through gene *N* and rightwards from promoter p_R through the *cro* gene (Fig. 2). Most of these initial transcripts terminate at sites t_{L1} and t_{R1}, immediately beyond genes *N* and *cro* respectively. The termination process at t_{L1} and t_{R1} is not always effective, but the few per cent of transcripts that escape here are terminated at subsequent sites, such as t_{R2} preceding gene *Q*. The early leftward transcript is translated to yield the *N* protein, which exerts its controlling effect by influencing RNA polymerase to ignore transcription termination signals (Adhya *et al.*, 1974; Franklin, 1974; Segawa and Imamoto, 1974).

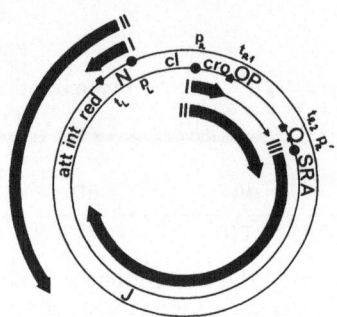

Figure 2   Transcription of lambda gene during lytic growth. I: Transcription immediately on induction or infection. In the absence of lambda's N protein, leftward transcription terminates largely at t_L, while rightward transcription terminates inefficiently at t_{R1}, beyond the *cro* gene. About 20% of the right-ward transcripts (thin arrow) escape termination at t_{R1} and proceed through genes O and P to be stopped at t_{R2}. II: After expression of the N gene transcription is influenced to ignore termination sites t_L, t_{R1} and t_{R2}. Leftward transcription leads to expression of the recombination (*red*) and integration (*int*) functions: rightward transcription proceeds through the O, P and Q genes, leading to DNA replication and the activation of late transcription. III: The product of the Q gene activates late transcription from promoter P_R to proceed through genes S and R, governing cell-lysis, and through genes A to J, coding for structural proteins of the phage head and tail.

In the presence of N protein, leftward transcription initiated at p_L is elongated through the *gam*, *red*, *xis* and *int* genes, while rightward transcription from p_R proceeds through the O and P genes, leading to the initiation of phage DNA replication, and through t_{R2} and gene Q. The product of the latter gene activates transcription from the late promoter, p_R', through genes S, R and the genes A to J in the circular phage chromosome.

The products of the regulatory genes N and Q are normally essential for phage development and plaque formation. Since the transcription-terminating site t_{R1} is inefficient, an N^- phage can express the O and P genes and replicate its DNA. If the terminator site t_{R2} is removed genetically, an N-defective phage can then express gene Q and proceed through late gene expression to maturity and plaque formation. Many lambdoid vectors contain a deletion of t_{R2}, *nin*5 (Court and Sato, 1969), that conveniently generates extra capacity for cloned DNA within the vector and makes the phage N-independent.

The product of the *cro* gene, after a delay of about ten minutes required to achieve the critical concentration, interacts directly with the template DNA at the complex p_L and p_R control sites and severely depresses the rate of the now superfluous transcription from these promoters (Radding and Shreffler, 1966; Pero, 1971; Echols *et al.*, 1973).

III   Replication and maturation of lambda DNA

A   DNA structure

Bacteriophage lambda DNA is replicated bidirectionally (Schnös and Inman, 1970) from an origin within the *O* gene (Denniston-Thompson *et al.*, 1977), in a process that requires the products of phage genes *O* and *P* acting in concert with host replication functions. During the first ten minutes or so after infection the phage genomes are replicated as monomeric circular molecules, but the rolling circle mode ensues to produce the multimeric concatenates that are the substrates for packaging of linear monomeric genomes during the phage maturation process.

B   DNA packaging and size selection

The lambda DNA packaging reaction imposes a stringent requirement for DNA size, only molecules between 78% and 105% of the wild-type genome length being packaged into viable particles (Weil *et al.*, 1973). While this property necessarily sets an upper limit on the size of DNA fragment that can be cloned in a lambda vector, this limit is about 22 kb, since the essential phage genes occupy only about 30 kb of lambda DNA.

Because there is both a lower and an upper limit to the size of a lambdoid phage genome, different strategies are required for the cloning of small versus large DNA fragments. A small fragment of DNA is most readily cloned into an INSERTION VECTOR, that has a single target for a restriction enzyme in a non-essential part of its genome. Ideally, the genome of the vector will be small enough to allow insertion of DNA fragments of up to several kilobase pairs, and the target for the restriction enzyme will be in such a location as to give insertional inactivation of a gene and easy recognition of recombinants by their changed phenotype. DNA fragments of more than 10 kb can be cloned into a REPLACEMENT VECTOR, allowing the replacement of a non-essential fragment of vector DNA between two widely spaced targets for a particular restriction enzyme. The size-selection concept demands that the non-essential vector fragment be either retained or replaced, but not simply deleted. The presence of the dispensable fragment can often be detected by a simple phenotypic test or genetic selection, allowing recombinants to be clearly distinguished or selectively isolated from the parental vector.

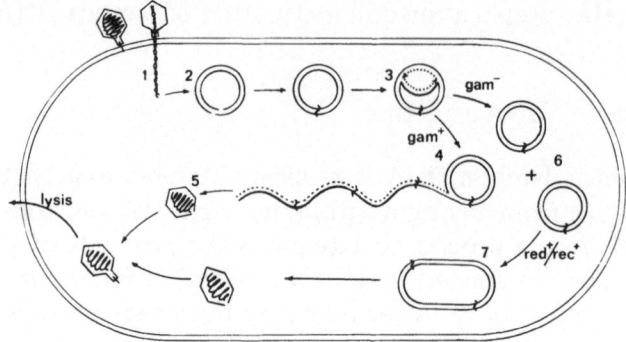

Figure 3   Replication and packaging of lambda DNA. Following phage adsorption the linear DNA is injected (1) and rapidly circularizes via its cohesive ends (2). The genome replicates as a monomeric circle (3), before the formation of concatenates via the rolling circular form (4). Packaging into phage heads normally occurs by excision of linear monomers from the concatenate (5). The formation of rolling circles is sensitive to the host's exonuclease V and is potentiated by the action of the *gam* gene product, an inhibitor of the nuclease. Phages lacking the *gam* gene can form packageable structures by recombination of monomeric circular forms (6) to give multimeric forms (7), catalysed by the phage *red* or the host *rec* recombination system.

It is worth noting that each replacement vector places a lower as well as an upper limit on the size of DNA fragment that can be accommodated.

C   Maturation of recombinant phages

The replication of lambda DNA switches fairly early in infection from the "theta" mode, duplicating monomeric circles, to the "rolling circle" form that makes the multimeric concatenates that are the maturable DNA species. Intermediates in this switching process are sensitive to attack by a host cell nuclease, exonuclease V (Mackay and Linn, 1974, 1976), the product of the *E. coli recB* and *recC* genes (Barbour and Clark, 1970). Replicating lambda DNA is protected against exonuclease V by the product of the phage's *gam* gene, a protein made early in infection that complexes with the nuclease and inhibits its action (Sakaki *et al.*, 1973).

The *gam* gene of phage lambda is located in the non-essential central region of the genome (Fig. 1), and recombinants made with several replacement vectors will have lost this gene. Such recombinants will be unable to make concatameric DNA via rolling circles, the formation of which will remain susceptible to the ravages of exonuclease V. These phages can make maturable forms of DNA by recombination of monomeric molecules into multimeric circular forms, from which linear monomers can be excised and packaged. The required recombination events can be catalysed by the *E. coli* recombination enzymes or by the phage-coded generalized recombination system. Since the *red* genes of the phage that govern generalized recombination are adjacent to the *gam* gene (Fig. 1), they will also tend to be missing from recombinants made with replacement vectors. Recombinants lacking both the *red* and *gam* genes will therefore be dependent on the *E. coli* recombination system for the generation of maturable forms of DNA. Lambda DNA is a poor substrate for the normal, *recA*-dependent recombination pathway of *E. coli*, but can be considerably improved by acquisition of a *chi*-mutation (cross-over *hot*-spot instigator) (Lam *et al.*, 1974) that generates a recombinogenic nucleotide sequence in the phage genome (Lam *et al.*, 1974; Henderson and Weil, 1975). Phages lacking the *red* and *gam* genes can also be propagated on *recB* or *recC* mutants of *E. coli* that do not produce exonuclease V. This stratagem has the advantage of allowing phage propagation under non-recombinogenic conditions, thereby minimizing rearrangement of cloned sequences, but tends to produce relatively low yields of phage.

D   *In vitro* packaging of phage DNA

The packaging of linear lambda genomes into mature phage particles can be achieved *in vitro* (Becker and Gold, 1975), using concentrated cell extracts made from induced lysogens containing prophages with mutations in genes coding for different steps in phage morphogenesis. When extracts from two lysogenic strains harbouring prophages with complementing capsid defects are mixed, endogenous and exogenous phage DNA is packaged into plaque-forming particles. In contrast to the behaviour *in vivo*, *in vitro* packaging will utilize linear monomeric DNA molecules, so that the efficiencies of packaging extracts can readily be determined using mature λ DNA extracted from phage heads (Hohn, 1975). Values of 10^7-10^8 phage/μg of λ DNA are usually obtained by established *in vitro* packaging procedures (Hohn and Murray, 1977; Sternberg *et al.*, 1977).

The strains used to prepare the packaging extracts carry prophages with mutations in their *cI* genes that make them readily heat-inducible. Nonsense mutations in capsid genes *A* and *E* (Sternberg *et al.*, 1977) or *D* and *E* (Hohn and Murray, 1977) prevent phage maturation *in vivo*. The prophages also carry mutations in gene *S*, a lysis function, so that the induced lysogens fail to lyse spontaneously and phage products continue accumulating for several hours beyond the normal time of lysis. This property is important in allowing the induced but intact cells to be harvested and concentrated before lysis. The *b*2 deletion present in the prophages inactivates the phage attachment site and blocks excision of prophage DNA after induction. The induced *b*2 prophage DNA replicates *in situ* in the host chromosome, giving amplification of late gene expression, but is not available for packaging into mature virions. To eliminate *in vitro* recombination of exogenous and endogenous phage DNAs, the major host recombination system is inactivated by a *recA* mutation and the generalized phage recombination system is eliminated by a lesion in the *redB* gene.

In vitro packaging under some conditions can be markedly size-selective, and this property can be used to enrich for recombinants generated from a vector with a small genome (Sternberg *et al.*, 1977). The selectivity can be eliminated by altering either the packaging conditions (Faber *et al.*, 1978) or the way in which the extracts are prepared (Hohn, 1979).

The enhanced efficiency of *in vitro* packaging gives this method a crucial advantage over the transfection procedures. It is also often of great expedience to be able to harvest recombinant phages without passage through host cells: relatively feeble recombinants will not be overgrown in the population, and host strains for subsequent manipulations can be chosen without regard for the ill-understood property of competence for DNA uptake.

Detailed discussions of both the theoretical and the practical aspects of *in vitro* packaging procedures have recently been published (Enquist and Sternberg, 1979; Hohn, 1979).

IV   The recognition of recombinant phages

In many situations, and especially when a particular donor DNA is in short supply, it is helpful to be able to distinguish recombinant phages from the parental vector. In favourable circumstances recombinants containing a gene of interest can readily be detected by restoration of a genetic defect in a mutant strain of *E. coli*, by direct detection of a gene product or by use of a suitable hybridization probe. In the general case, however, this is not possible and phage vectors have been designed that allow the ready recognition of recombinants by loss, through inactivation or replacement, of genes affecting the phenotype of the phage.

A   Direct screens

The plaques of lambdoid phages are characteristically turbid, due to the growth within the area of the plaque of bacterial cells that have undergone the lysogenic response to phage infection. Crucial to the turbidity of the plaque is the synthesis of the phage repressor, the protein that negatively controls the expression of other phage genes and renders a lysogenic cell immune to infection by phages of the same immunity-specificity. Insertion vectors have been developed that have their unique target sites within the *cI* gene that codes for the phage repressor. Recombinant phages carry insertions in their *cI* genes and are readily distinguished by their clear plaques.

Insertion vectors of this type rely on the fact that the *cI* gene of lambdoid phage 434 carries single targets for EcoRI and HindIII. Elimination of the targets from other regions of the genomes of λimm434 hybrid phages has produced a series of vector phages for use with one of these two enzymes (Murray and Murray, 1975; Blattner *et al.*, 1977; Murray *et al.*, 1977). The Charon 7 vector lacks both EcoRI and HindIII targets outside the immunity region, and can therefore be used with either of these enzymes or with the combination, when heterologously terminated fragments can be cloned (Blattner *et al.*, 1977).

The immunity-insertion vectors are very convenient to use for cloning EcoRI- or HindIII-generated fragments of DNA of up to about 10 kb. Recombinants generally grow well, and are more vigorous than the parental vector because of the advantages of both the clear plaque phenotype and the more suitable size of the genome. Although these recombinants cannot themselves be used to lysogenize or transduce a sensitive *E. coli* host, they *can* be integrated into a recombination-proficient lysogen carrying a prophage of ϕ434 immunity with relatively high efficiency. Genes inserted into the ϕ434 *cI* gene can potentially be expressed by leftward transcription from the phage promoters *pre* and *prm*. This situation can be effectively exploited by infection of a UV-irradiated homo-immune lysogen, when the only significant gene expression arises from transcription from *prm* or from a cloned promoter. The use of ^{35}S-methionine during infection allows the products of the cloned genes to be detected on autoradiographs following polyacrylamide gel electrophoresis (Jaskunas *et al.*, 1977).

Phages carrying the *lacZ* gene of *E. coli* provide a convenient detection system when used in conjunction with appropriate indicator plates. The most effective indicator is 5-bromo-4-chloro-3-indolyl-β-D-galactoside (XG), which is hydrolysed by β-galactosidase to produce a deep-blue, non-diffusing pigment, indigo blue. A vector carrying the intact *lacZ* gene will produce blue plaques on plates containing XG, on either a Lac$^+$ or a Lac$^-$ host strain. Recombinants carrying DNA inserted at either the EcoRI site or the SstI site in *lacZ* will give colourless plaques on a host carrying a deletion of the *lacZ* gene (Blattner *et al.*, 1977; Pourcel and Tiollais, 1977).

The blue plaque visual screen is also applicable in conjunction with λ*lac* replacement vectors designed for use with EcoRI. Since the *lacZ* gene itself contains an EcoRI target, inversion of the central fragment destroys the *lacZ* gene and would produce a Lac⁻ plaque on a host strain carrying a deletion of the *lacZ* gene. However, replacement of the central fragment can be distinguished from inversion by plating on a *lacZ*⁺ host on plates containing XG. Irrespective of orientation, the multiple copies of the *lac* operator produced as the phage DNA replicates are sufficient to titrate the host's *lac* repressor and induce β-galactosidase synthesis from the chromosomal copy of the *lacZ* gene, resulting in a blue plaque (Blattner *et al.*, 1977). Alternatively, retention of the central fragment in either orientation can be detected because it gives rise to β-galactosidase activity by allelic complementation with appropriate, omega-donor mutants (Ullmann *et al.*, 1967) of the host *lacZ* gene (Murray *et al.*, 1977).

Other replacement vectors allowing the use of *lac* indicator technology are those that include *amber*-suppressor genes in their filler fragments. Vectors containing the *supE* (SuII) gene can be used with EcoRI, while those carrying *supF* (SuIII) allow cloning with HindIII (Murray *et al.*, 1977). In both cases, vectors give blue plaques and recombinants colourless ones on a host strain with an *amber* mutation in *lacZ*.

Replacement vectors carrying the *E. coli araB*, *A* and *D* genes on an exchangeable central fragment allow recombinants to be distinguished by plating on an Ara⁻ host strain on arabinose—MacConkey indicator plates (Struhl and Davis, 1977).

B   Indirect screens

Several vectors allow the detection of recombinant phages by simple spot-tests or replica plating procedures. These are generally more laborious than the direct visual screens, but several hundred phages can usually be tested in an hour, and a preliminary screen of this kind can be very useful in eliminating subsequent work requiring a more difficult screening procedure, or in merely ascertaining the frequency of successful cloning events.

Insertion into the EcoRI site *srI*λ3 inactivates the phage's generalized recombination system, and the *red*⁻ recombinants can be detected by their failure to grow on *pol*A or ligase-deficient host strains (Murray and Murray, 1974). In practice, the test involves picking individual plaques following transfection or infection of a normal host and testing them for growth on a control strain and a *pol*A (or *ligt. s.*) host strain. (In general, *red*⁻ phages tend to be less vigorous and to make smaller plaques than *red*⁺ phages, so that the screening can be biased in favour of recombinant phages by choosing the smaller plaques.)

Several of the available lambdoid vectors contain genes from either the *bio* or the *trp* operons of *E. coli*. Inactivation or replacement of these *E. coli* genes can be detected by testing phages from individual plaques for their ability to "complement" a *bio*A or appropriate *trp* defect in an *E. coli* host strain. The tests are carried out on lawns of mutant cells plated in the absence of the required nutrient. A positive response shows as growth of transductant colonies or, if the phage is incapable of making repressor, by an area of lysis surrounded by a halo of cell growth. In tests such as these it is important to avoid growth of bacterial colonies that may be carried over from the original indicator plate. This can often be achieved by careful choice of host strain, including the use of antibiotic-resistant mutants in the presence of the antibiotic for the final step.

The sensitivity of phage particles to disruption by chelating agents such as EDTA or pyrophosphate increases markedly with genome size (Parkinson and Huskey, 1971). It is usually possible with an insertion vector to find a concentration of chelating agent that will discriminate between the parental phage and recombinants carrying a relatively small insert. Simple tests using plates containing pyrophosphate can be sensitive enough to detect an insert of 1 kb of DNA.

A change in genome size results in an alteration in the buoyant density of the phage particle, and this can be detected very sensitively by centrifugation to equilibrium in a caesium chloride gradient. This method can be used to provide a physical selection for recombinant phages with either increased or decreased genome sizes (Sternberg *et al.*, 1977; Philippsen *et al.*, 1978).

C   Positive selection for recombinant phages

Several methods are now available for preventing either the re-formation or the propagation of vector genomes while selectively allowing growth of recombinant phages.

A method of general application with replacement vectors is to purify the essential fragments of restricted phage DNA away from the replaceable central fragment (Thomas *et al.*, 1974). This is readily achieved by sedimentation through a sucrose density gradient, after having annealed the two terminal fragments via the natural cohesive ends of the lambda genome (Maniatis *et al.*, 1978). The ease of purifying the vector "arms" can often be enhanced by treatment of the vector DNA with a second restriction endonuclease that attacks only the central fragment. The purified fragments of vector DNA require a further "filler" fragment in order to produce a viable phage and therefore provide a strong selection for recombinants with a certain minimal genome size.

A similar size-selection can be achieved by use of a special *pel*⁻ host strain that inhibits the *penetration of l*ambda DNA (Scandella and Arber, 1974). This host effectively prevents the growth of phages with reduced genome sizes, preferentially allowing propagation of phages with a full complement of DNA (Emmons *et al.*, 1975). A *pel*⁻ host strain can be used to select for the cloning of DNA fragments of a chosen size, the value of which depends on the DNA content of the particular vector phage (Cameron *et al.*, 1977; Philippsen *et al.*, 1978).

1   *The Spi*⁻ *phenotype as a selection for recombinant phages*

Wild-type λ does not form plaques on *E. coli* strains carrying the P2 prophage (Lederberg, 1957), a phenotype known as Spi⁺ for sensitive to *P2* interference. Spi⁻ mutants of λ that can form plaques on P2 lysogens have simultaneously lost the function of the *red* and *gam* genes (Lindahl *et al.*, 1970; Zissler *et al.*, 1971). Since the *red* and *gam* genes of phage λ are located close to each other in the non-essential region of the phage genome, it is possible to arrange for these genes to be replaced by foreign DNA and therefore to select for recombinant phages by their acquisition of the Spi⁻ phenotype. The full Spi⁻ phenotype requires the presence in the phage of a *chi* site to stimulate the formation of maturable, multimeric DNA molecules via the host recombination system (Lam *et al.*, 1974; Henderson and Weil, 1975: see section III.C). Existing vectors allowing the selection of recombinants by growth on a P2-lysogenic host can be used in conjunction with *Eco* RI (Murray and Murray, 1975), *Hind*III (Murray and Murray, 1975), the *Bam*HI-related enzymes (Karn *et al.*, 1980; Rimm *et al.*, 1981; N.E. Murray, pers. comm.), or all three of these (Loenen and Brammar, 1980).

It is worthy of note that Spi⁻ phages cannot be grown on *rec*A⁻ host strains (Lindahl *et al.*, 1970), where they fail to make packageable molecules (Lam *et al.*, 1974). In addition, because of the absence of the normally inhibitory *gam* gene product, the host's recombination system will be unusually active during infection of a *rec*⁺ host. If it is considered important to minimize recombinational rerrangement of cloned DNA sequences, as it generally will be when eukaryotic DNA is being studied, the best available compromise is to use a *recB, C*⁻ host strain. Detailed analysis of mammalian globin genes in Spi⁻ recombinant phages failed to reveal any rearrangements after propagation in a *recB, C*⁻ host (Jeffreys *et al.*, 1982).

A powerful selection for recombinant phages can be achieved with the replacement vector λgtWES.T5-622, for use with EcoRI, constructed by Davison *et al.* (1979). The central region of this vector contains tandem copies of a 1.8 kb fragment of phage T5 DNA that includes gene *A*3. The product of this gene causes abortion of infection on hosts carrying the plasmid ColIb (Lanni, 1969). The combination of a special host strain and size-selection ensures that all plaques are due to phages carrying fragments of donor DNA.

D   Hybridization screening

Phages containing particular cloned DNA sequences can readily be located by nucleic acid hybridization whenever a suitable radio-labelled RNA or DNA probe is available. The original colony hybridization procedure of Grunstein and Hogness (1975) is readily adapted for use with plaques picked onto bacterial lawns grown on nitrocellulose filters (Kramer *et al.*, 1976), or with the DNA of individual plaques transferred directly from the agar plate onto a filter (Benton and Davis, 1977). A variation of the latter procedure includes an amplification step, propagating the transferred phages on a lawn of bacteria on the filter before *in situ* hybridization (Woo, 1979), that decreases the time required for autoradiography.

Since phage plaques contain up to 10^7 phage particles as well as an equivalent quantity of unpackaged phage DNA the sensitivity of the plaque-hybridization procedure is high. Phages containing unique sequences from a mammalian genome can readily be detected on 9-cm diameter plates carrying more than 10^5 very small plaques (Jeffreys *et al.*, 1982).

E   Screens dependent on gene expression

Several methods are available that allow the detection or selection of recombinant phages that are able to express a gene of interest in an infected host cell. By appropriate choice of the host strain and the vector phage it is sometimes possible to make the growth of the phage itself dependent on the product of the desired gene. The ligase genes of *E. coli* (Cameron *et al.*, 1975; Borck *et al.*, 1976) and of bacteriophage T4 (Wilson and Murray, 1979) have been cloned into λ vectors by making use of the fact that *red*⁻ phages will not grow on ligase-deficient hosts. Phages carrying genes that are expressed to produce a product that compensates for a defect in an *E. coli* host strain give rise to plaques surrounded by a halo of bacterial growth on plates lacking the desired metabolite (Franklin, 1971). This technique has allowed the isolation of phages carrying biosynthetic and catabolic functions from *E. coli* (Borck *et al.*, 1976) and functions that are not normally present in that host (Drew *et al.*, 1980). Phages containing a gene coding for a DNA modification activity have been selected by their insensitivity to the cognate restriction enzyme (Borck *et al.*, 1976; Sain and Murray, 1980).

Successful selection often depends on the formation of stable, lysogenic transductants, with the phage genome integrated into that of its host. This is most efficiently achieved when the recombinant phage remains integration-proficient and directs the integration of its own genome at the specific attachment site on the host chromosome. Many vectors give rise to integration-defective recombinants, due to loss or inactivation of *att* or the *int* gene, while others cannot give rise to lysogens because of defective *cI* genes. In these cases it is possible to form lysogens, and therefore selectable transductants, by use of an integration-proficient, cI^+ helper phage, or by infection of lysogens carrying prophages of similar immunity (Struhl *et al.*, 1976). The use of the helper phage tends to give higher transduction frequencies, and has the extra advantage that it allows the use of $recA^-$ hosts, thereby stabilizing the tandem dilysogen.

In the special case when the recombinant phages carry chromosomal DNA from *E. coli*, the cloned fragment directs integration directly into the homologous region of the host chromosome, catalysed by the host's recombination system but still requiring a functional *cI* gene (Shimada *et al.*, 1972; Schrenk and Weisberg, 1975). This property can be turned to advantage to transfer mutations from the *E. coli* chromosome onto a transducing phage, or to allow extension of the cloned segment of DNA following aberrant excision from the prophage state.

The products of genes that are expressed within infected cells can be detected in individual plaques by staining directly for enzymic activity (Brammar *et al.*, 1980) or by immunological techniques. The simplest immunochemical detection methods use *in situ* immunoassays on phage plaques, visualizing positive clones by an immunoprecipitin reaction in the agar (Sanzey *et al.*, 1976; Skalka and Shapiro, 1976). Variations using radio-iodinated antibody or *Staphylococcus aureus* protein A have increased the sensitivity of the method to the extent that it will detect one antigen molecule per *E. coli* cell (Broome and Gilbert, 1978; Ehrlich *et al.*, 1978). A method that is particularly suitable for screening for rare antigen-positive plaques depends on the incorporation of a horseradish peroxidase/antibody conjugate in the top-layer agar of a plaque-assay plate (Kaplan *et al.*, 1981). The antigen produced during plaque formation precipitates the peroxidase/antibody complex locally, and is detected using a standard chromogenic assay for peroxidase activity. Positive plaques, indicated by the local brown pigment, can readily be distinguished even amongst a background of more than 10^6 plaques per 9-cm petri plate.

The immunochemical approach has the advantage that it will detect enzymically inactive proteins, including fusion peptides, provided they retain immunological activity.

V   The expression of genes cloned into phage lambda

By making use of the extensive background knowlege of the transcription of lambda genes and its control, it is possible to obtain high rates of expression of genes cloned in a lambda vector. The particular strategem to employ depends on whether or not the cloned gene carries with it a promoter that functions efficiently in *E. coli*.

A   Genes with their own promoter

The expression of a gene cloned together with a functional promoter can readily be amplified by taking advantage of the increase in copy number when the phage DNA replicates. The inclusion of a S^- mutation in the phage delays cell lysis and allows prolonged DNA replication and gene expression (Harris *et al.*, 1967; Müller-Hill *et al.*, 1968). Studies using phages containing the *trp* genes of *E. coli* have shown that Q^- phages are considerably more effective than S^- derivatives for obtaining high yields of gene products (Hopkins *et al.*, 1976; Moir and Brammar, 1976). The Q^- mutants of lambda are defective in expression of their late genes (Dove, 1966; Couturier *et al.*, 1973), so that they not only fail to lyse the host cell but also fail to remove phage genomes from the pool of transcribable DNA by packaging into phage heads.

The efficacy of this approach to amplify expression of a gene from its own promoter has been amply confirmed in studies with phages carrying the *lig* (DNA ligase: Panasenko *et al.*, 1977), *pol*A (DNA polymerase I: Murray and Kelley, 1979) and *dna*Z (DNA polymerase III: γ subunit: Hubscher and Kornberg, 1980) genes of *E. coli*. In practice it is most convenient if the recombinant phage is integration-proficient, so that it can be carried as a normal prophage before being induced to replicate and excise by a temperature shift. (It must also carry a temperature-sensitive mutation in the *cI* gene, of course.) Phages that are integration-defective can lysogenize via host-catalysed recombination, either through genetic homology between the cloned genes and the *E. coli* chromosome or, if the cloned DNA is non-homologous, through the homology with a pre-formed, lambdoid prophage.

Both these circumstances allow successful amplification of gene expression on induction, though the yields of gene products are not so high as in the ideal case (N. E. Murray and W. J. Brammar, unpublished). Lytic infection at moderately high multiplicities of infection is also very effective on the laboratory scale.

For analytical purposes, the expression of genes from their own promoters can readily be monitored following infection of a UV-irradiated, homo-immune lysogenic host strain. The resident phage-repressor blocks transcription of all the genes of the superinfecting phage except *cI* itself and any that are transcribed from their own promoters. The proteins made in such an infection can be labelled with ^{35}S-methionine, separated by SDS polyacrylamide gel electrophoresis and visualized by autoradiography (Jaskunas *et al.*, 1977).

B   Expression from P_L

The lambda promoter p_L, governing early leftward transcription of the phage genes *N* to *int*, has several features that make it attractive for controlling the expression of cloned genes. It is a very powerful promoter (Franklin, 1971; Davison *et al.*, 1974) whose activity can be conveniently regulated by manipulation of the activity of a temperature-sensitive repressor. In addition, a nucleotide sequence *nutL* (Rosenberg *et al.*, 1978; Salstrom and Szybalski, 1978), just downstream from the promoter, allows the lambda *N* gene product to act as a powerful antagonist of transcription termination (Adhya *et al.*, 1974; Franklin, 1974). Segawa and Imamoto, 1974). The activity of p_L is also controlled by the product of the lambda *cro* gene, acting to decrease the rate of leftward transcription early in infection (Sly *et al.*, 1971). A *cro* mutation can be used to give enhanced and prolonged transcription from p_L (Franklin, 1971; Sly *et al.*, 1971), but the high copy number that is necessary to obtain dramatic yields of gene products is very difficult to achieve with *cro*⁻ phages, which show enfeebled DNA replication (Moir and Brammar, 1976).

Studies with the *polA* gene of *E. coli* cloned in a lambda vector (Murray and Kelley, 1979) showed that the p_L promoter could still be effective in a *cro*⁺ phage, where efficient DNA replication can compensate for the much lower rate of leftward transcription.

An important feature of transcription from p_L is that it can very effectively compete with, and sometimes totally block, convergent rightward transcription of a gene cloned into the *N* operon in the rightward orientation (Hopkins *et al.*, 1976; Ward and Murray, 1979). It can therefore be important to ensure that a gene is in the *l* orientation, even when transcription is expected to emanate largely from a cloned promoter.

C   Expression from other phage promoters

During attempts to optimize the expression of the gene for the DNA ligase of bacteriophage T4 cloned into lambdoid vectors, the most effective amplification was obtained via transcription initiated at the lambda late promoter, p'_R (Murray *et al.*, 1979). Because transcription from p'_R requires the *Q* gene product, the λT4*lig* phage was made conditionally defective in the gene, *E*, that codes for the major capsid protein (Casjens *et al.*, 1970), to block DNA packaging and prolong transcription, as well as in gene *S* to prevent cell lysis. In this case the T4*lig* gene had been cloned, using EcoRI, into a λ*lac* vector, λNM816, to give an integration-proficient recombinant phage lacking the section between 40% and 54% on the lambda map (Wilson and Murray, 1979). Although this region of the lambda genome contains sequences that block rightward transcription from p'_R (Krell *et al.*, 1972), none is located to the left of the *Hind*III target at 48.5% (Burt and Brammar, 1982). Thus genes cloned in the appropriate orientation at *srI*λ1 (44%) or *shnd*λ1 (48%) should be capable of high rates of transcription from p'_R. Genes located further rightwards at *shnd*λ3 (56.4%) are not effectively expressed from p'_R, even when the phage carries a deletion from *srI*λ1 to *srI*λ2 (44% to 54%) (Hopkins *et al.*, 1976). Genes in this position are capable of being expressed at a very low rate in the prophage (Hopkins *et al.*, 1976), presumably from a weak promoter for rightward transcription in the *b*2 region.

D   General considerations

The effective utilization of phages for amplification of the expression of cloned genes depends on the presence of appropriate mutations in the phage genome. In general, the *Q*-mutation is effective in conjunction with genes being expressed from their own promoter or from p_L, while the combination of *S* and *E* gene defects is expedient to harness expression from p'_R. The combination of an integration- and excision-proficient phage with a heat-inducible defect in *cI* gives the most convenient system for amplifying expression. Although vectors containing useful mutations have been constructed, it is simpler in practice to introduce the desired mutations by phage crosses, having first cloned the gene(s) of interest into a robust and easily used vector. Characterization of the initial recombinant phage often allows subcloning into a more convenient vector for gene expression.

The lambda vectors afford a potential advantage over the commonly used plasmid systems for expressing cloned genes in allowing strict but flexible control of both promoter activity and gene copy number. The cloned gene can be maintained at a single copy per chromosome in the prophage state, before being triggered into expression from a phage promoter and replication to very high dosage by a simple temperature shift. In addition, following either induction or lytic infection, there is no requirement for long-term survival of the host cell, so that even genes specifying potentially lethal products can be employed.

VI   Constructing gene banks with lambda vectors

The construction of genomic libraries, or "gene-banks", by cloning fragments of complex genomes has been covered in detail (Dahl *et al.*, this series, Vol. 2). Phage vectors are proving particularly useful for this purpose, and derivatives that are especially suitable for making libraries have recently been described.

The number of clones required to contain an entire genome obviously depends on the size of that genome and the average size of the DNA fragments cloned. The probability (P) that a given unique sequence is present in a collection of N clones is given by

$$P = 1 - (1-f)^N$$

where f is the average fraction of the genome represented by one clone.

Thus the size of the collection required is

$$N = \frac{\ln(1-P)}{\ln(1-f)}$$

(Clarke and Carbon, 1976). Thus, for a mammalian genome of about 3×10^6 kb, assuming an average cloned fragment of 15 kb, a 99% probability of obtaining a given sequence would necessitate about 10^6 recombinants.

The large number of clones required for a complete library of a complex genome makes efficient cloning and recovery of recombinants of paramount importance. It is largely for this reason that phage or cosmid systems, allowing recovery of recombinants by *in vitro* DNA packaging, are preferred to simple plasmids as vectors for constructing genomic libraries of higher eukaryotes.

Lambdoid vectors used for preparing genomic libraries are generally replacement vectors that will accommodate relatively large fragments of donor DNA. Maniatis *et al.* (1978) used the

EcoRI/replacement vector, Charon 4A (Blattner *et al.*, 1979) to prepare libraries from *Drosophila*, silkmoth and rabbit genomes. Blunt-ended fragments of donor DNAs were prepared both by shearing followed by digestion with nuclease S1 and by partial digestion with restriction enzymes that make frequent breaks and produce blunt ends. DNA fragments were fractionated by sucrose gradient centrifugation, and those of about 20 kb were subsequently cloned into Charon 4A using EcoRI linker sequences. Since the synthetic linkers have to be cleaved by EcoRI to generate cohesive ends, any EcoRI targets with the donor DNA fragments must first be made resistant to the EcoRI enzyme. This is achieved by methylation with the EcoRI modification-methylase in the presence of *S*-adenosyl methionine (Greene *et al.*, 1975). The left and right arms of the Charon 4A vector DNA were annealed through the lambda cohesive ends before restriction of the phage DNA with EcoRI, and the internal fragments of the vector genome were removed by sucrose gradient centrifugation. This step both increases the efficiency of formation of recombinant genomes and minimises the number of non-recombinant phages recovered. The donor and vector DNA fragments were ligated at high DNA concentrations to favour the formation of concatameric recombinants that are the preferred substrates for *in vitro* DNA packaging.

Analysis of the relative efficiencies of the two procedures showed that partial digestion was 3—16 times more efficient than using shear. The best experiments yielded 5.6×10^5 plaques/μg of eukaryotic DNA, with insert sizes averaging 19 kb.

The phage libraries prepared by *in vitro* packaging were amplified about 10^6-fold by preparation of low density plate lysates. When the amplified libraries were compared to donor DNA preparations by Cot analysis, no change in the complement of single copy sequences could be detected. All three libraries were successfully screened with gene-specific hybridization probes, and the frequencies with which recombinants were located suggest both that the libraries were complete and that the detection of specific recombinants by hybridization (Benton and Davis, 1977) was efficient.

The cloning of blunt-ended fragments of DNA using synthetic linkers, though validated by the successful experience of Maniatis' group, involves potentially inefficient steps that could be avoided by use of fragments with cohesive ends. Partial digests made with EcoRI or HindIII can be cloned into an appropriate replacement vector, but this approach has the disadvantage that the enzymes make relatively few breaks, so that the ends of the fragments will not be sufficiently randomly placed. The use of the EcoRI endonuclease under sub-optimal conditions, when it loses specificity to become a tetranucleotide-recognizing enzyme while still producing the AATT cohesive end (Polisky *et al.*, 1975), is one approach to this problem. A second, generally more useful approach has been opened up by the development of lambdoid vectors for use with the BamHI endonuclease.

The lambda genome contains five targets for BamHI, one of which, *sbh*IλI, is within an essential gene (Perricaudet and Tiollais, 1975; Haggerty and Schlief, 1976). Phages that had lost *sbh*IλI by mutation were isolated by successive enrichment for genomes that were resistant to the *in vitro* action of BamHI (Klein and Murray, 1979; Rimm *et al.*, 1980). Once this target had been removed, it was relatively easy to generate insertion or replacement vectors for use with BamHI. Such phages are of especial interest as vectors, since they can be used to clone not only fragments generated by BamHI (recognition sequence G$^\downarrow$GATCC), but those produced by any enzyme that generates the 5'-GATC cohesive end, including BclI (T$^\downarrow$GATCA), BglII (A$^\downarrow$GATCT), XhoII (Pu$^\downarrow$GATCPy) and Sau3A ($^\downarrow$GATC). Partial digestion with the latter enzyme is an effective means of generating an approximately random population of high molecular weight DNA fragments and is therefore finding use in the construction of genomic libraries.

Several replacement vectors for use with the BamHI-related series of enzymes have been described. Because of the positions of the BamHI targets in the lambda genome, it is easy to arrange for the replacement of a central fragment that includes the *red* and *gam* genes of the vector, generating recombinants that can be distinguished or selected by their Spi$^-$ phenotype (see section IV.C).

The vector λL47 (Loenen and Brammar, 1980) has a replaceable, central fragment bounded by BamHI-targets *shbh*Iλ3 (58.1%) and *sbh*Iλ4 (71.3%), with a further 8 kb of cloning capacity provided by two deletions. It will accommodate 5'-GATC-ended fragments of 4.7—19.6 kb. Inversion of the BamHI-generated central fragment of this vector separates the *red* and *gam* genes from their normal promoter, p$_L$, but derivatives with the inverted fragment express their *red* and *gam* genes from an alternative promoter and retain the Spi$^+$ phenotype (Loenen and Brammar, 1980). Growth on a P2-lysogen, the Spi$^-$ phenotype, can therefore be used as a valid selection for recombinant phages. The λL47 vector can also be used in a similar way as a replacement vector for fragments generated by EcoRI or HindIII, for which it has a somewhat larger capacity. Its value in the construction of genomic libraries has recently be established by Jeffreys *et al.* (1982), who have used it with BamHI and Sau3A to generate 10^6 recombinants/μg of brown lemur DNA.

A vector of different design but similar concept to λL47 has been described by Karn *et al.* (1980). The central fragment of this vector, λ1059, has been manipulated in such a way that it includes the leftward promoter p_L. Inversion of this fragment therefore does not separate *red* and *gam* from their promoter and does not alter the Spi⁺ phenotype. The λ1059 vector will accommodate DNA fragments in the range 6—24 kb. Karn *et al.* (1980) have used this vector to clone size-selected Sau3A-digested fragments of nematode DNA. Without physical removal of the central fragment of the vector DNA, selecting recombinants on a P2-lysogenic host strain, they obtained 2×10^5 — 5×10^5 recombinant phages/μg of nematode DNA.

Rimm *et al.* (1980) have described vectors in the Charon series that carry BamHI cloning sites. The one that is most useful for constructing genomic libraries, Charon 30, contains two tandem copies of a central "stuffer" fragment, and has a cloning capacity of 6—19 kb. Like the other BamHI/replacement vectors it allows selection of recombinants by Spi⁻ phenotype. Charon 30 is also a vector for fragments generated by HindIII and EcoRI, and can be used with EcoRI and BamHI in combination to clone fragments with heterologous ends.

It is possible to use vectors like λL47, λ1059 or λ Charon 30 to generate genomic libraries without pre-sizing the donor DNA digest, since the DNA packaging system selects inserts of the appropriate size. This practice is not to be recommended, however, since it can lead to artefacts due to the ligation of two or more normally noncontiguous fragments of donor DNA. The selection of a population of donor fragments of a size close to that required to fill the capacity of the vector should minimize this problem. Treatment of the donor DNA fragments with alkaline phosphatase or calf intestinal phosphatase should also prevent self-ligation (Ullrich *et al.*, 1977), but often tends to decrease the yield of recombinants.

Scalenghe *et al.* (1981) have used the insertion vector λ641 (Murray *et al.*, 1979) and *in vitro* DNA packaging to clone restriction fragments from a specific region of a polytene chromsome from the salivary gland of *Drosphila melanogaster*. The region of interest was dissected by micromanipulation from a squashed preparation of salivary gland of *Drosophila melanogaster*. The region of interest was ligated to an excess of vector DNA in nanolitre volumes in an oil chamber. After *in vitro* DNA packaging, 80 recombinant phages were recovered from about 10 pg of *Drosophila* DNA. Preliminary analysis showed that the DNA from sample clones hybridized to the appropriate region of the *Drosophila* genome. Clones obtained in this way from polytene chromosomes can be used to isolate recombinants from a conventional genomic library to give saturation coverage of a region of interest.

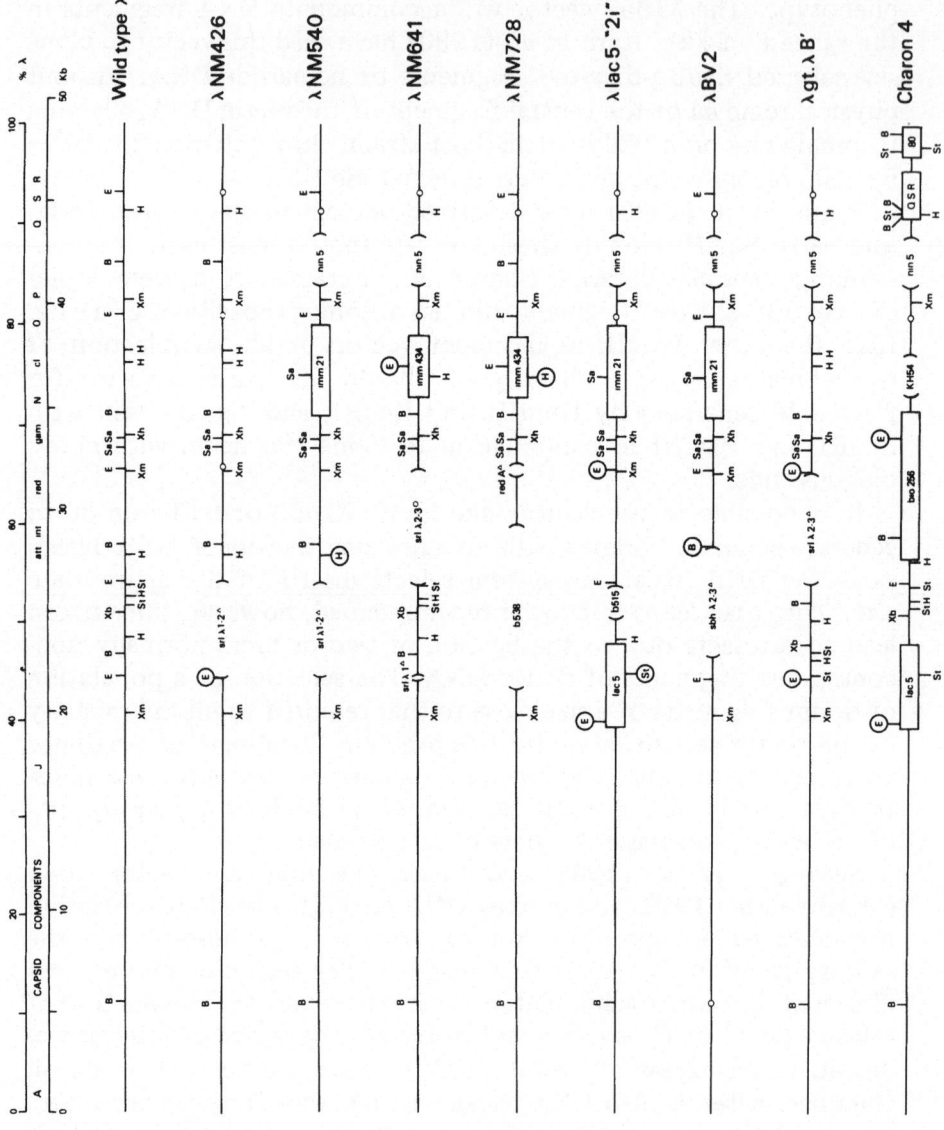

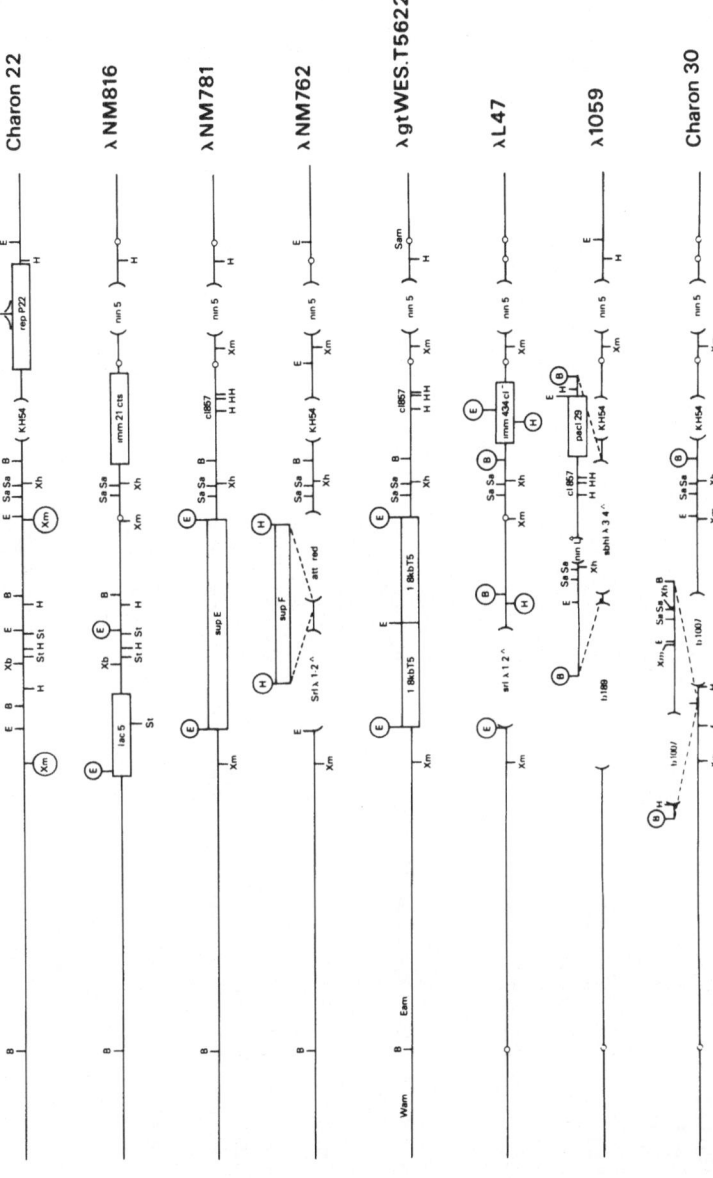

Figure 4 Physical maps of selected lambda vectors. More than 100 different lambda vectors have been described. Those presented here are chosen for a particular property or because they have featured prominently in the literature. For a more comprehensive list of lambda vectors see the review by Williams and Blattner (1980). B = *Bam*HI; E = *Eco*RI; H = *Hind*III; Sa = *Sal*I; St = *Sst*I; Xh = *Xho*I; Xm = *Xma*I. symbols enclosed in circles show the preferred cloning sites for each vector.

Table 1 Capacities of some lambda vectors.

Phage	Enzyme	Insert size (kb)	Recognition of recombinants	Comments on recombinants	Reference
λNM426	EcoRI	0–6.6		Int$^+$, cI$^+$,	Murray and Murray (1975)
λNM540	HindIII	0–11.8	—	Int$^+$, cI$^+$	Murray and Murray (1975)
λNM641	EcoRI	0–11.6	clear plaques	Red$^-$ (see also Charons 6 and 7)	Murray et al. (1977)
λNM728	HindIII	0–12.2	clear plaques	Red$^-$ (see also Charon 7)	Murray et al. (1977)
λlac5-2i	EcoRI	6.2–19.2	Lac$^-$(a)		
λBV2	SstI	0–8.5	Lac$^-$(b)	Int$^+$, Red$^+$, cIts	Pourcel and Tiollais (1977)
	BamHI	0–12.6	—	Int$^-$, cI$^+$ (cI$^-$ version = λNM1068) (see also Charon 27 [Rimm et al. 1980])	Klein and Murray (1979)
λgt.λB′	EcoRI	2.1–15.1	—	Int$^-$, Red$^-$	Thomas et al. (1974)
Charon 4	EcoRI	7.0–20.0	Lac$^-$(a), Bio$^-$	(see also λgt.λB′WES, Enquist et al., 1976 and λgtZJvir.λB′, Donogue and Sharp, 1977) Int$^-$, Red$^-$, Gam$^-$, (will not grow on RecA$^-$ host) (see also Charon 4A, and A$^-$; B$^-$ version)	Blattner et al. (1977)
Charon 22	XmaI	3.8–16.8	—	Int$^-$, Red$^-$	Williams and Blattner (1980)
λNM816	EcoRI	6.3–19.3	Lac$^-$(c)	Int$^+$, Red$^+$, cIts	Wilson and Murray (1979)
λNM781	EcoRI	2.2–15.2	Lac$^-$(c)	Int$^-$, Red$^-$, cIts	Murray et al. (1977)
λNM762	HindIII	2.6–15.6	Lac$^-$(c)	Int$^-$, Red$^-$	Murray et al. (1977)
λgt WES. T5/622	EcoRI	2.1–15.1	ColIb-insensitive (d)	Int$^-$, Red$^-$, cIts, W$^-$, E$^-$ S$^-$	Davison et al. (1979)
λL47	EcoRI	8.4–21.4	Spi$^-$(e)	Int$^-$, Red$^-$, Gam$^-$, N$^-$, cI$^-$ (will not grow on recA host)	Loenen and Brammar (1980)
	HindIII	6.8–19.8	Spi$^-$(e)		
	BamHI	4.2–17.2	Spi$^-$(e)	Int$^-$, Red$^-$, Gam$^-$, N$^+$, cI$^-$ (will not grow on recA host) (cI 857 version available: D. W. Burt, University of Leicester)	

Vector	Enzyme				Reference
λ1059	BamHI	8.0–21.0	Spi⁻(e)	Int⁻, Red⁻, Gam⁻ (will not grow on recA⁻ host)	Karn et al. (1980)
λCharon 30	EcoRI	4.5–17.5	—	Int⁻, Red⁻	Rimm et al. (1980)
	HindIII	0–11.7	—	Int⁻, Red⁺	
	BamHI	6.1–19.1	Spi⁻(e)		
	EcoRI⁺			Int⁻, Red⁻, Gam⁻	
	BamHI	7.2–20.2	Spi⁻(e)	(will not grow on recA host)	

Capacities of some lambdoid vectors

Wild type λ DNA has single targets for XbaI (51.0) and XhoI (69.0), so that any phage with a smaller than wild-type genome is potentially a vector for use with these enzymes. In addition, the two SalI targets at 67.7 and 68.8% and the two SstI targets at 51.7 and 53.9% are in non-essential regions of the genome. Many of the vectors allow the use of XbaI, XhoI, SalI or SstI in conjunction with another enzyme such as EcoRI, HindIII or BamHI; such cases will be evident from inspection of the restriction maps in Fig. 4.

Insert sizes have been calculated assuming a lower limit to the genome size of 78% (38.3 kb) and an upper limit of 105% (51.6 kb) lambda wild-type length.

[a]The vector gives blue plaques (Lac⁺ phenotype) on a lac⁺ host on plates containing 5-bromo-4-chloro 3-indolyl-β-D-galactoside (XG), a non-inducing substrate of β-galactosidase, due to titration of the host's lac repressor by the lac operator on the phage DNA. Recombinants, having lost the lac operator, give colourless plaques under these conditions.

[b]The vector gives blue (Lac⁺) plaques on a lacZ⁻ host strain on plates containing XG, due to the β-galactosidase encoded by the phage. Recombinants with an insertion in the lacZ gene give colourless (Lac⁻) plaques under these conditions.

[c]Vectors carrying the supE or supF amber suppressor genes from E. coli give Lac⁺ (blue) plaques on a host carrying a lacZ amber mutation of plates containing XG and IPTG as inducer of the chromosomal lac operon. Recombinant phages lack the suppressor gene and give colourless plaques on these plates.

[d]The λgt.T5/622 vector cannot grow on a host strain carrying colIb, which selects against the T5-A3 gene on the central fragments.

[e]Recombinants can be recognized or selected by their ability to grow on a P2-lysogenic host that prevents growth of the vector phage.

VII Conclusions and future developments

Lambda vectors suitable for many different purposes are now available, and the techniques have been described for using these vectors to clone specific DNA fragments, construct genomic libraries, analyse gene expression or amplify the yield of a gene product. The attraction of lambda as a cloning vehicle lies in the versatility that derives from the fund of knowledge of the physiology and genetics of the lambdoid phages, and in the relative ease of propagation and purification of phage DNA.

Lambda vectors are used almost exclusively for the cloning of DNA fragments with cohesive ends, and vector development has centred round manipulation of the targets for EcoRI, HindIII and BamHI. Derivatives are available, however, that allow cloning of fragments generated by SalI (CTCGAG) and XmaI (CCCGGG) (see Fig. 4).

Now that the precedent has been established for the elimination of target sites by *in vitro* enrichment (Klein and Murray, 1979; Rimm *et al.*, 1980), it should be possible to remove further such sites from the lambda genome. Generally useful vectors could then be established by the introduction of synthesic DNA sequences containing an array of suitable cloning sites. The construction of both an insertion vector and a replacement vector along these lines would allow the cloning of both small and relatively large DNA fragments with a wide range of restriction enzymes (Rimm *et al.*, 1980).

The upper limit of about 25 kb on the size of a DNA fragment that can be cloned into a lambda vector is imposed by the phage's DNA packaging system and the presence of the many essential genes in the phage genome. In principle, it should be possible to construct a special host strain of *E. coli* that harbours the essential late genes of lambda and could be induced to express them. Such a host would allow most of the vector genome to be replaced by donor DNA and thereby greatly increase the cloning capacity of the lambdoid vectors.

The vast genetic potential of the lambdoid vector systems has been little exploited to date. The ability of lambda phages carrying *E. coli* DNA to recombine with the *E. coli* chromosome has been used to potentiate clone extension (Hopkins *et al.*, 1976). This approach should not be limited to the analysis of *E. coli* genes, and it is likely that the ability of hybrid phages to recombine with each other, or with chimaeric plasmids, through the homology provided by overlapping fragments of cloned DNA will shortly provide the basis of a genetic method for clone extension and chromosome walking.

VIII Acknowledgements

I am grateful to Fred Blattner, John Davison, Philippe Kourilsky, Noreen Murray, Waclaw Szybalski and Pierre Tiollais for providing published and unpublished information, and to David Burt and Wilhelmine Loenen for valuable discussion.

IX References

Adhya, S., Gottesman, M. and de Crombrugghe, B. (1974). *Proc. Natn. Acad. Sci. U.S.A.* **71**, 2534–2538.

Barbour, S. D. and Clark, A. J. (1970). *Proc. Natn. Acad. Sci. U.S.A.* **65**, 955–961.

Becker, A. and Gold, M. (1975). *Proc. Natn. Acad. Sci. U.S.A.* **73**, 4174–4178.

Blattner, F. R., Williams, B. G., Blechl, A. E., Denniston-Thompson, K., Faber, H. E., Furlong, L-A., Grunwald, D. J., Kiefer, D. O., Moore, D. D., Schumm, J. W., Sheldon, E. L. and Smithies, O. (1977). *Science, N. Y.* **197**, 161–169.

Benton, W. D. and Davis, R. W. (1977). *Science, N. Y.* **196**, 180–182.

Borck, K., Beggs, J. D., Brammar, W. J., Hopkins, A. S. and Murray, N. E. (1976). *Molec. Gen. Genet.* **146**, 199–207.

Brammar, W. J., Muir, S. and McMorris, A. (1980). *Molec. Gen. Genet.* **178**, 217–224.

Brooks, K. (1965). *Virology* **26**, 489–499.

Broome, S. and Gilbert, W. (1978). *Proc. Natn. Acad. Sci. U.S.A.* **75**, 2746–2749.

Burt, D. W. and Brammar, W. J. (1982). *Mol. Gen. Genet.*, in press.

Cameron, J. R., Philippsen, P. and Davis, R. W. (1977). *Nucl. Acids Res.* **4**, 1429–1448.

Campbell, A. and Del Campillo-Campbell, A. (1963). *J. Bacteriol.* **85**, 1202–1207.

Casjens, S., Hohn, T. and Kaiser, A. D. (1970). *Virology* **42**, 496–507.

Clarke, L. and Carbon, J. (1976). *Proc. Natn. Acad. Sci. U.S.A.* **72**, 4361–4365.

Court, D. and Sato, K. (1969). *Virology* **39**, 348–352.

Couturier, M., Dambly, C. and Thomas, R. (1973). *Molec. Gen. Genet.* **120**, 231–252.

Davison, J., Brammar, W. J. and Brunel, F. (1974). *Molec. Gen. Genet.* **130**, 9–20.

Davison, J., Brunel, F. and Merchez, M. (1979). *Gene* **8**, 69–80.

Denniston-Thompson, K., Moore, D. D., Kruger, K. E., Furth, M. E. and Blattner, F. R. (1977). *Science, N. Y.* **198**, 1051–1056.

Dove, W. (1966). *J. Molec. Biol.* **19**, 187–201.

Drew, R. E., Clarke, P. H. and Brammar, W. J. (1980). *Molec. Gen. Genet.* **177**, 311–320.

Echols, H., Green, L., Oppenheim, A. B., Oppenheim, A. and Honigman, A. (1973). *J. Molec. Biol.* **80**, 203–216.

Ehrlich, H. A., Cohen, S. N. and McDevitt, H. O. (1978). *Cell* **13**, 681–689.

Emmons, S. W., MacCosham, V. and Baldwin, R. L. (1975). *J. Molec. Biol.* **91**, 133–146.

Enquist, L. and Sternberg, N. (1979). *In* "Methods in Enzymology" (Ed. R. Wu) Vol. **68**, 281–298. Academic Press, New York.

Faber, H., Kiefer, D. and Blattner, F. (1978). Quoted in Enquist and Sternberg (1979).

Franklin, N. C. (1971). *In* "The Bacteriophage Lambda" (Ed. A. D. Hershey) 621—638. Cold Spring Harbor Laboratory, Cold Spring Harbor.

Franklin, N. C. (1974). *J. Molec. Biol.* **89**, 33—48.

Greene, P. J., Poonian, M. S., Nussbaum, A. L., Tobias, L., Garfen, D. E., Boyer, H. W. and Goodman, H. M. (1975). *J. Molec. Biol.* **99**, 237—261.

Grunstein, M. and Hogness, D. S. (1975). *Proc. Natn. Acad. Sci. U.S.A.* **72**, 3961—3965.

Haggerty, D. M. and Schlief, R. F. (1976). *J. Virol.* **18**, 659—663.

Harris, A. W., Mount, D. W. A., Fuerst, C. R. and Siminovitch, L. (1967). *Virology* **32**, 553—569.

Henderson, D. A. and Weil, J. (1975). *Genetics* **79**, 143—174.

Hendrix, R. W. (1971). *In* "The Bacteriophage Lambda" (Ed. A. D. Hershey) 355—370. Cold Spring Harbor Laboratory, Cold Spring Harbor.

Hohn, B. (1975). *J. Molec. Biol.* **98**, 93—106.

Hohn, B. (1979). *In* "Methods in Enzymology" (Ed. R. Wu) Vol. 68, 299—309. Academic Press, New York.

Hohn, B. and Murray, K. (1977). *Proc. Natn. Acad. Sci. U.S.A.* **74**, 3259—3263.

Hopkins, A. S., Murray, N. E. and Brammar, W. J. (1976). *J. Molec. Biol.* **107**, 549—569.

Hubscher, U. and Kornberg, A. (1980). *J. Biol. Chem.* **255**, 11698—11703.

Jaskunas, S. R., Fallon, A. M., Nomura, M., Williams, B. G. and Blattner, F. R. (1977). *J. Biol. Chem.* **252**, 7355—7364.

Jeffreys, A. J., Barrie, P. A., Harris, S., Fawcett, D. H., Nugent, Z. J. and Boyd, A. C. (1982). *J. Molec. Biol.*, in press.

Joyner, A., Isaacs, L. N. Echols, H. and Sly, W. (1966). *J. Molec. Biol.* **19**, 174—186.

Kaplan, D. A., Naumovski, L. and Collier, R. J. (1981). *Gene* **13**, 211—220.

Karn, J., Brenner, S., Barnett, L. and Cesareni, G. (1980). *Proc. Natn. Acad. Sci. U.S.A.* **77**, 5172—5176.

Klein, B. and Murray, K. (1979). *J. Molec. Biol.* **133**, 289—294.

Kramer, R. A., Cameron, J. R. and Davis, R. W. (1976). *Cell* **8**, 227—232.

Krell, K., Gottesman, M. E. and Parks, J. S. (1972). *J. Molec. Biol.* **68**, 69—82.

Lam, S. T., Stahl, M. M., McMilin, K. D. and Stahl, F. W. (1974). *Genetics* **77**, 425—433.

Lanni, Y. T. (1969). *J. Molec. Biol.* **44**, 173—184.

Lederberg, S. (1957). *Virology* **3**, 496—

Lindahl, G., Sironi, G., Bialy, H., Calendar, R. (1970). *Proc. Natn. Acad. Sci. U.S.A.* **66**, 587—594.

Loenen, W. A. M. and Brammar, W. J. (1980). *Gene* **10**, 249—259.

MacKay, V. and Linn, S. (1974). *J. Biol. Chem.* **249**, 4286—4294.

MacKay, V. and Linn, S. (1976). *J. Biol. Chem.* **251**, 3716—3719.

Mamatis, T., Hardison, R. C., Lacy, E., Lamer, J., O'Connell, C., Quorn, D., Sim, G. K. and Efstratiadis, A. (1978). *Cell*, **15**, 687—701.

Moir, A. and Brammar, W. J. (1976). *Molec. Gen. Genet.* **149**, 87—99.

Müller-Hill, B., Crapo, L. and Gilbert, W. (1968). *Proc. Natn. Acad. Sci. U.S.A.* **59**, 1259—1264.

Murray, K. and Murray, N. E. (1975). *J. Molec. Biol.* **98**, 551—564.

Murray, N. E. and Kelley, W. S. (1979). *Molec. Gen. Genet.* **175**, 77—87.

Murray, N. E. and Murray, K. (1974). *Nature, Lond.* **251**, 476—481.

Murray, N. E., Brammar, W. J. and Murray, K. (1977). *Molec. Gen. Genet.* **150**, 53—61.

Murray, N. E., Bruce, S. A. and Murray, K. (1979). *J. Molec. Biol.* **132**, 493—505.

Panasenko, S. N., Cameron, J. R., Davis, R. W. and Lehman, I. R. (1977). *Science, N. Y.* **196**, 188–189.

Parkinson, J. S. and Huskey, R. J. (1971). *J. Molec. Biol.* **56**, 369–384.

Pero, J. (1971). *In* "The Bacteriophage Lambda" (Ed. A. D. Hershey), 599–608, Cold Spring Harbor Laboratory, Cold Spring Harbor.

Pericaudet, M. and Tiollais, P. (1975). *FEBS Lett.* **56**, 7–11.

Philippsen, P., Kramer, R. A. and Davis, R. W. (1978). *J. Molec. Biol.* **123**, 371–386.

Polisky, B., Greene, P., Garfin, D. E., McCarthy, B. J., Goodman, H. M. and Boyer, H. W. (1975). *Proc. Natn. Acad. Sci. U.S.A.* **72**, 3310–3314.

Pourcel, C. and Tiollais, P. (1977). *Gene* **1**, 281–286.

Ptashne, M. (1971). *In* "The Bacteriophage Lambda" (Ed. A. D. Hershey) 221–237. Cold Spring Harbor Laboratory, Cold Spring Harbor.

Radding, C. M. and Shreffler, D. C. (1966). *J. Molec. Biol.* **18**, 251–261.

Rimm, D., Horness, D., Kucera, J. and Blattner, F. R. (1980). *Gene* **12**, 301–309.

Rosenberg, M., Court, D., Shimatake, H., Brady, C. and Wulff, D. (1978). *In* "The Operon" (Eds J. H. Miller and W. S. Reznikoff) 345–371. Cold Spring Harbor Laboratory, Cold Spring Harbor.

Sain, B. and Murray, N. E. (1980). *Molec. Gen. Genet.* **180**, 35–46.

Sakaki Y., Karu, A. E., Linn, S. and Echols, H. (1973). *Proc. Natn. Acad. Sci. U.S.A.* **70**, 2215–2219.

Salstrom, J. S. and Szybalski, W. (1978). *J. Molec. Biol.* **124**, 195–221.

Sanzey, B., Mercereau, O., Ternynck, T. and Kourislky, P. (1976). *Proc. Natn. Acad. Sci. U.S.A.* **73**, 3394–3397.

Scalenghe, F., Turco, E., Edstrom, J. E., Pirrotta, V. and Melli, L. (1981). *Chromosoma* **82**, 205–216.

Scandella, D. and Arber, W. (1974). *Virology* **58**, 504–513.

Schnos, M. and Inman, R. B. (1970). *J. Molec. Biol.* **51**, 61–73.

Schrenk, W. J. and Weisberg, R. A. (1975). *Molec. Gen. Genet.* **137**, 101–107.

Segawa, T. and Imamoto, F. (1974). *J. Molec. Biol.* **87**, 741–754.

Shimada, K., Weisberg, R. A. and Gottesman, M. E. (1972). *J. Molec. Biol.* **63**, 483–503.

Skalka, A. and Shapiro, L. (1976). *Gene* **1**, 65–79.

Sly, W. S. Rabideau, K. and Kolber, A. (1971). *In* "The Bacteriophage Lambda" (Ed. A. D. Hershey) 575–588. Cold Spring Harbor Laboratory, Cold Spring Harbor.

Sternberg, N. Tiemeier, D. and Enquiest, L. (1977). *Gene* **1**, 255–280.

Struhl, K. and Davis, R. W. (1977). *Proc. Natn. Acad. Sci. U.S.A.* **74**, 5255–5259.

Struhl, K., Cameron, J. R. and Davis, R. W. (1976). *Proc. Natn. Acad. Sci. U.S.A.* **73**, 1471–1475.

Thomas, M., Cameron, J. R. and Davis, R. W. (1974). *Proc. Natn. Acad. Sci. U.S.A.* **71**, 4579–4583.

Ullmann, A., Jacob, F. and Monod, J. (1967). *J. Molec. Biol.* **24**, 339–343.

Ullrich, A., Shine, J., Chirgwin, J., Pictet, R., Tisher, E., Rutter, W. J. and Goodman, H. M. (1977). *Science, N. Y.* **196**, 1313–1316.

Ward, D. F. and Murray, N. E. (1979). *J. Molec. Biol.* **133**, 249–266.

Weil, J., Cunningham, R., Martin, R.III, Mitchell, E. and Bolling, B. (1973). *Virology* **50**, 373–380.

Wilson, G. and Murray, N. E. (1979). *J. Molec. Biol.* **132**, 471–491.

Woo, S. L. C. (1979). *In* "Methods in Enzymology" (Ed. R. Wu) Vol. 68, 389–395. Academic Press, New York.

Zissler, J. Signer E. R. and Schaefer, F. (1971). *In* "The Bacteriophage Lambda" (Ed. A. D. Hershey) 469–476. Cold Spring Harbor Laboratory, Cold Spring Harbor.

Expression of cloned genes in eukaryotic cells using vector systems derived from viral replicons

PETER W. J. RIGBY

Cancer Research Campaign Eukaryotic Molecular Genetics Research Group, Department of Biochemistry, Imperial College of Science and Technology, London SW72AZ, UK

I   Introduction

A   Why develop eukaryotic cloning systems?

The application of recombinant DNA techniques has revolutionized our understanding of the structure, organization and expression of eukaryotic genes. Well-established methodologies are available for the construction of clone libraries containing either cDNA copies of mRNAs (Williams, this series, Vol. 1) or genomic DNA (Dahl *et al.*, this series, Vol. 2) and any sequence can be isolated from such libraries if a method for screening is available. Once a suitable clone has been isolated, even if it contains only a small portion of the gene or genetic locus in question, the entire gene or locus, together with large amounts of flanking DNA, can be isolated by repeated screenings of appropriate libraries (chromosome walking) and the organization of these sequences can be studied in the source organism and in genetically related individuals and species using the transfer hybridization technique developed by Southern (1975). The availability of chemical DNA sequencing techniques (Maxam and Gilbert, 1980) and, more particularly, the development by Sanger and his colleagues of the extraordinarily rapid dideoxy sequencing method which employs cloning in M13 (Sanger, 1981), mean that the determination of the complete nucleotide sequence of large segments of cloned DNA is now routine. Finally, the adaptation of the original transfer hybridization technique to the detection of RNA (Alwine *et al.*, 1977; Thomas, 1980) and the establishment and refinement of the endonuclease Sl method for mapping the 5′ and 3′ termini of mRNA molecules and the location of introns (Berk and Sharp, 1977; Favaloro *et al.*, 1980) allow the pattern of transcription to be readily analysed if a cloned probe is available.

We can thus deduce an extremely detailed picture of the structure of a eukaryotic gene, of its adjacent sequences, which all believe contain the information controlling its expression, and of the nuclear and cytoplasmic RNA molecules transcribed from it. The successes of this approach, perhaps most notably the discovery of introns and RNA splicing and the elucidation of the somatic DNA rearrangements involved in the generation of antibody diversity, are striking, yet it has one fundamental weakness. It can not lead to a direct correlation between particular nucleotide sequences and defined biological functions. Inspection of nucleotide sequences has led to the identification of putative control signals. The sequence AAUAAA is found close to the 3′ end of polyadenylated mRNAs and was hypothesized to be a signal for transcriptional termination and/or polyadenylation (Proudfoot and Brownlee, 1976). More recently, the sequence

TATA^A/$_T$A^T/$_A$ or a variant thereof, has been identified approximately 30 nucleotides upstream of the initiation site of most eukaryotic RNA polymerase II transcription units (Goldberg, 1978; Corden *et al.*, 1980) and has therefore been implicated in the initiation of transcription. However, direct proof of the function of motifs identified by sequence comparison can only be achieved by introducing mutations into them and then assessing the functional capability of the mutants.

The techniques for cloning and manipulating eukaryotic genes in *Escherichia coli* are not adequate for this purpose. The fundamental mechanisms of gene expression differ markedly between prokaryotic and eukaryotic organisms and even in those cases where the power of recombinant DNA technology has been used to ensure the expression of eukaryotic genes in *E. coli* (Carey, this series, in preparation) the mechanisms controlling that expression are those of *E. coli*. The construction of a bacterial strain in which a eukaryotic gene responds appropriately to a eukaryotic control stimulus, e.g. a steroid hormone, is probably impossible. Even if such a technical *tour de force* could be achieved, the results obtained from such a heterologous system would be open to the severest of criticisms. What is required is the development of systems by which cloned eukaryotic genes, and mutated derivatives, can be reintroduced into a eukaryotic environment so that their expression can be analysed.

One approach to this problem has been the establishment of *in vitro* transcription systems derived from eukaryotic cells (Weil *et al.*, 1979; Manley *et al.*, 1979). While such systems have allowed the definition of some of the properties of eukaryotic promoters, the extrapolation of information gained from them to *in vivo* situations is not clear. Mutations in the "TATA" box have different effects when assayed *in vivo* and *in vitro* (Benoist and Chambon, 1981; Mathis and Chambon, 1981). More importantly, these systems are not, in general, subject to the regulation of gene expression observed *in vivo*; globin genes are transcribed in cell-free systems derived from HeLa cells which do not express globin (Luse and Roeder, 1980). While some cases of regulation have been described, transcription from the Simian virus 40 (SV40) early promoter is appropriately regulated by large T-antigen (Rio *et al.*, 1980) and a RNA polymerase I *in vitro* system transcribes rRNA genes only when the extract is prepared from actively growing cells (Grummt, 1981), it seems unlikely that such systems, in which the template for transcription is naked DNA, will in the immediate future allow the reconstruction of complex regulatory phenomena. The microinjection of cloned DNA into *Xenopus laevis* oocytes has also been used to assay the effects of defined mutations on transcription. In this system the injected DNA is assembled into chromatin but once again correct regulation has not been demonstrated. The characteristics and uses of the *in vitro* and *Xenopus* oocyte transcription systems have been reviewed in this series (Wickens and Laskey, this series, Vol. 1).

To circumvent these problems it is necessary to develop systems for the reintroduction of cloned genes into, in the first instance, cultured eukaryotic cells and, in the final analysis, into eukaryotic organisms. Ideally such systems should be totally general, they should allow the gene to be introduced into all cell types. This requirement precludes the generalized use of viruses with a restricted host range. Moreover, it is necessary to develop selection procedures, analogous to those employed in prokaryotic cloning, which allow the recovery of those cells which receive DNA or, alternatively, to develop gene transfer techniques so efficient that all of the cells in a population take up the DNA.

Such techniques will overcome many of the problems associated with *in vitro* transcription systems and with heterologous systems such as oocyte injection. Cloned DNA introduced into cultured animal cells is accurately transcribed and thus the sequences controlling the initiation of transcription can be analysed. Moreover, correct polyadenylation and splicing occur in cells opening the way for the study of the signals specifying these post-transcriptional processing events. The RNA polymerase II *in vitro* transcription systems commonly used are restricted to initiation; correct termination and/or polyadenylation have not been documented. *In vitro* systems for the splicing of polymerase II transcripts have been described (Weingartner and Keller, 1981; Goldenberg and Raskas, 1981) but their major use is likely to be as assay systems for the purification of the cellular components that mediate splicing, rather than for the assessment of the effect of mutations on the splicing process. A major goal of much current research in eukaryotic molecular biology is the elucidation of the mechanisms involved in the induction of transcription by extracellular stimuli, e.g. hormones, and in the regulation of tissue-specific and developmental-stage-specific gene expression. While the *in vitro* reconstitution of such processes remains our ultimate goal, the basic features of such processes must first be defined and this can only be achieved by reintroducing cloned genes into an appropriate cellular environment.

In all of these cases the experimental approach depends upon the introduction of defined mutations into putative regulatory regions and the subsequent analysis of the biological consequences of the mutation. The procedures for *in vitro* mutagenesis are beyond the scope of this review. Suffice it to say that entire restriction fragments can be removed (Lai and Nathans, 1974), small deletions can be introduced by controlled exonucleolytic digestion (Carbon *et al.*, 1975; Myers and Tjian, 1980) and point mutations can be introduced by direct chemical mutagenesis (Shortle and Nathans, 1978) or by enzymatic repair mechanisms (Weissmann *et al.*, 1979; Weber *et al.*, 1981). More recently, the rapid development of techniques for the chemical synthesis of oligonucleotides has led to their use in the induction of defined mutations (Hutchison *et al.*, 1978; Wallace *et al.*, 1979; Gillam *et al.*, 1980) and one can expect an ever-increasing application of this methodology.

A further ramification of the ability to express cloned genes in cultured cells derives from the fact that such cells are endowed with a multitude of biosynthetic capabilities not possessed by *E. coli.* The production of large amounts of eukaryotic polypeptides which require post-translational modification, e.g. glycosylation or phosphorylation, for their biological activity will depend upon the application of eukaryotic cloning systems. Similar arguments apply to the use of recombinant DNA techniques to analyse the biosynthesis of complex sub-cellular structures.

The tremendous technical power and facility of *E. coli* cloning techniques will remain a prerequisite for the isolation and characterization of eukaryotic genes. Indeed, in many of the cases I shall discuss not only the initial cloning but also the subsequent manipulations, constructions and mutagenic steps have been performed in *E. coli*, with the eukaryotic cloning system being used only for the final analytical step. Nevertheless, the development of powerful, technically tractable and general methodologies for cloning in eukaryotic cells is an absolute requirement if our ever-increasing knowledge of the structure and organization of eukaryotic genes is to be translated into a detailed appreciation of how they are expressed and how that expression is controlled.

B   What eukaryotic cloning systems are available?

A variety of approaches has been employed to reintroduce cloned genes into eukaryotic cells. One tactic is to use a lower eukaryote that can be grown and manipulated with the same ease as *E. coli.* The obvious candidate is the yeast, *Saccharomyces cerevisiae*, which has been intensively studied genetically and biochemically. While the development of yeast cloning systems (Beggs, this series, Vol. 2) has led to tremendous developments in our understanding of the molecular biology of yeast itself, the use of yeast as a heterologous host for genes from other eukaryotes has been somewhat unrewarding. Hall and his colleagues used genetic complementation of a yeast auxotroph to clone a gene from *Drosophila melanogaster* (Henikoff *et al.*, 1981) and have recently demonstrated the efficient expression in yeast of a human interferon gene (Hitzeman *et al.*, 1981). Interferon genes do not, however, contain introns, and in the only reported case in which the expression of an intron-containing heterologous gene has been analysed in yeast, the rabbit β-globin gene was not correctly transcribed (Beggs *et al.*, 1980).

A second method is to use co-transformation with the thymidine kinase (tk) gene of herpes simplex virus (HSV). tk⁻ cells can not grow in selective HAT medium; if the HSV tk gene is introduced into such cells by calcium phosphate co-precipitation (Graham and van der Eb, 1973) the genetic defect can be overcome and tk⁺ cells can be recovered by growth in the selective medium. Wigler *et al.* (1979) showed that during this procedure not only the tk DNA, but also any other DNA that is covalently ligated to it or merely mixed with it, is taken up and integrated into the chromosomes of the host cell. Such non-selected sequences are frequently expressed (Mantei *et al.*, 1979; Breathnach *et al.*, 1980). A review of this system has recently appeared (Scangos and Ruddle, 1981) and it will be discussed in a forthcoming article in this series (Wilkie, in preparation). For the purposes of this review it is important to note the following points. The uptake and integration of the exogenous DNA occur via a complex pathway involving large concatameric structures and extensive rearrangement of the transforming DNA can occur. In many cell lines derived in this way only one or two copies of the exogenous DNA are present. While this is advantageous in some applications, low copy number is not desirable when the objective is analysis of primary mechanisms such as transcriptional initiation or splicing. However, the major disadvantage of this technique is that it can only be applied when tk⁻ cells are available and the generation of tk⁻ mutants of the specialized cell types one would like to study is a difficult and time-consuming process.

An alternative to the tk system is simply to introduce the cloned gene by calcium phosphate co-precipitation and then to analyse for its expression several days later without applying a selection for cells which have successfully taken up DNA. The problem with this approach is that only 1—2% of the cells take up DNA when the original co-precipitation procedure is used, and thus the level of expression is often below the sensitivity of even the most powerful immunological and biochemical methods. However, subtle alterations in the technical procedure now allow a ten-fold increase in the number of cells that can be induced to take up exogenous DNA (Chu and Sharp, 1981). More importantly, Schaffner (1980) showed that cloned DNA can be transferred directly from *E. coli* to cultured cells. In his procedure bacterial cells carrying the cloned DNA are converted to protoplasts and then fused to cultured cells using the polyethylene glycol method originally developed for use in somatic cell genetics. Schaffner reported that when the bacteria used carried a plasmid containing SV40 DNA and the recipient cells were permissive hosts for this virus, up to 7% of the cells could be shown immunologically to be expressing SV40 T-antigen. Rassoulzadegan *et al.* (1982) have refined Schaffner's procedure so that it is possible to transfer DNA to all of the cells in a culture. The power of this technique is obvious. It eliminates the need even to isolate DNA and allows every cell in the culture to express the cloned gene. A further development in this direction has been the demonstration that sequences from SV40 DNA can enhance the expression of eukaryotic genes to which they are linked (Banerji *et al.*, 1981). The properties of these enhancer sequences will be discussed later in this review. However, it is clear that these two developments applied in concert will greatly simplify the analysis of eukaryotic gene expression. The procedures can be applied to any cell type that can be grown in culture, they do not require mutant cells or selective systems and they are technically straightforward. The disadvantage is that they do not lead to the establishment of stable cell lines carrying the gene of interest and thus the DNA transfer process must be repeated for each set of experiments.

However, the most obvious way to clone genes in eukaryotic cells is to apply exactly the same rationales used so successfully with *E. coli.* Cloning in *E. coli* involves the use of either plasmid (Thompson, this volume) or bacteriophage (Brammar, this volume) vectors which have the capacity to replicate autonomously in *E. coli* and which can be selected for in some way. Endogenous plasmids have not been recognized in higher eukaryotic cells, although a more detailed analysis of the small, polydisperse circular DNA found in many cultured cells (DeLap *et al.*, 1978) might prove informative in this regard. Moreover, it has recently become apparent that some viruses replicate episomally. The major effort in the construction of eukaryotic vectors has, however, been in the adaptation of viruses to serve as generalized transducers in the same ways that λ and the filamentous phages such as M13 have been so successfully tailored to act as vectors in *E. coli.* It is with the use of animal viruses as vectors that this article will be primarily concerned.

II Criteria for the design of animal virus vectors

The considerations to be taken into account when designing animal virus vectors are closely analogous to those which have guided the design of bacteriophage λ vectors (Brammar, this volume). The simplest way to use a viral genome as a vector is to identify a region of that genome which is not essential for lytic growth and then to replace this non-essential region with foreign DNA. Unfortunately, those animal viruses which have been characterized in sufficient detail to be considered as vectors do not have the large non-essential region that has been so effectively exploited in λ. Therefore in most of the cases to be discussed the foreign DNA replaces essential gene(s) and the recombinant genome is thus absolutely defective and must be propagated in the presence of a helper virus. However, in some cases the helper viral gene(s) can be incorporated into the genome of the host cell and this is likely to be an increasingly used tactic.

The size of recombinant DNA molecules which can be propagated as infectious virions is limited by the amount of DNA which can be packaged into a virus particle. This size limit is a more acute constraint than was originally thought because we now know that many eukaryotic genes are extremely large; e.g. the chicken proα-2-collagen gene is 38 kb (Wozney *et al.*, 1981) and the mouse dihydrofolate reductase (DHFR) gene is longer than 42 kb (Nunberg *et al.*, 1980). Thus far no *in vitro* packaging system has been developed for an animal virus and so even when the recombinants are to be propagated as virions the initial infection must be done with DNA.

The final necessity is that the transcriptional programme of the virus is clearly understood. This is much more important for animal virus vectors than for phage vectors. In most work with λ or M13 the viral genome is used simply to replicate the foreign DNA whereas in eukaryotic cloning experiments the objective is almost always the expression of the inserted gene. Moreover, in eukaryotic systems it is necessary to consider not only transcriptional initiation and termination signals but also the pattern of post-transcriptional splicing.

All eukaryotic vector systems which use virions to propagate the recombinant genome, with the notable exception of retroviruses, have one serious disadvantage: the multiplication of the virus kills the host cell. In many cases one would like to produce a stable cell line which continuously replicates and expresses the exogenous DNA while suffering no ill effects. Much attention has therefore been focussed on the development of systems in which the recombinant DNA replicates as an episome. Episomal systems have another great advantage in that they are not subject to packaging constraints; it should be possible to use them to replicate and express even the largest eukaryotic genes.

A final class of viral vector exploits the fact that all presently available systems are based on tumour viruses and thus the viral genome itself provides a selection mechanism by virtue of its ability to transform the growth properties of normal cells. Most tumour viruses integrate their DNA into the chromosomes of the host cell during transformation. Thus, so long as the insertion of foreign DNA does not inactivate the transforming gene, which in most cases is clearly identified, selection for the transformed phenotype should ensure that the foreign DNA is also covalently integrated. Such systems have not, however, found widespread use. Only a few copies of the transforming DNA are integrated so one does not obtain the amplification of gene expression which occurs during a productive infection. Moreover, integration is often accompanied by rearrangement events which can not be predicted or controlled.

It is clear that there are many possible approaches to the use of animal virus genomes as vector systems and the system of choice will vary according to the precise purposes of the experiment. Moreover, the development of this technology is presently very rapid and, while some profitable directions for future work can be clearly discerned, it is certain that the current rate of technical advance will render some systems obsolete even before they are fully developed. It is therefore necessary that those who contemplate the use of such systems adopt a very flexible attitude. With this caveat in mind I shall now review the various types of viral vector together with some examples of their use.

III   Systems for the propagation of recombinant DNA molecules as virions

A   Simian virus 40

1   Introduction

The virus which has received most attention as a possible vector for the propagation of exogenous genetic information in cultured eukaryotic cells is Simian virus 40 (SV40). This virus was an obvious choice as its genome is small, it was the first animal virus DNA to be completely sequenced and we have detailed knowledge of many aspects of its molecular biology. I plan here to describe only those facets of the virus directly relevant to its use as a vector. All aspects of the molecular biology of SV40 have recently been authoritatively reviewed (Tooze, 1980) while Martin (1981) has written an excellent review of cellular transformation by SV40 and Lebowitz and Weissman (1979) have discussed the transcription of the viral genome in considerable detail. Those planning to experiment with SV40 would be well-advised to consult these sources.

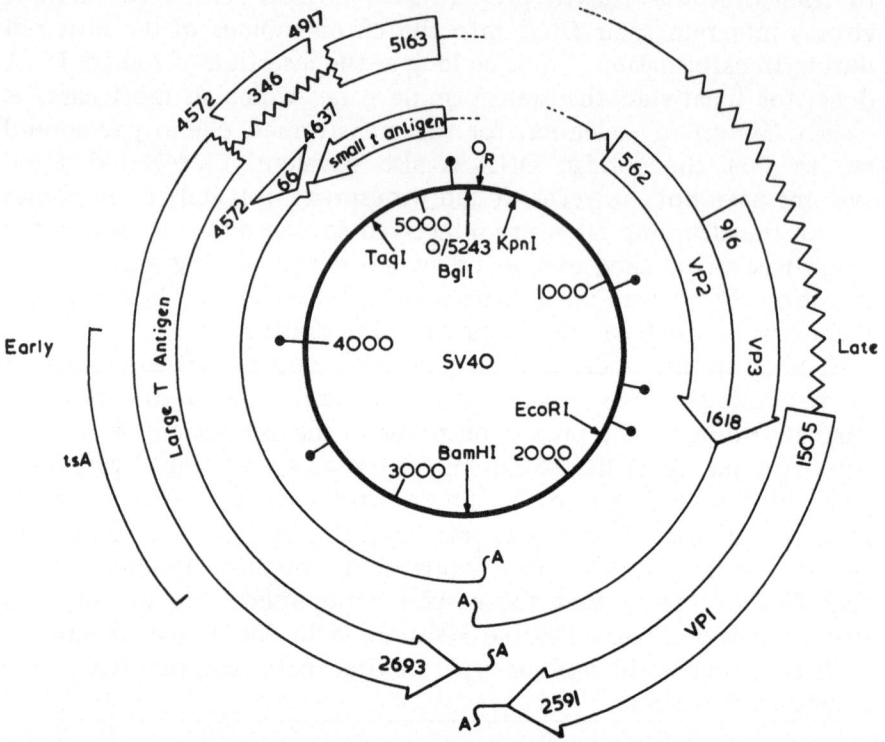

Figure 1   A functional map of the Simian virus 40 genome. The nucleotide numbering system of Buchman, Burnett and Berg (see Tooze, 1980) is used. Within the mRNAs: – – – – indicates the regions in which the 5′ termini map; ⋀⋁⋀⋁ indicates regions removed by splicing; ▭ indicates translated sequences. The numbers at the beginning and end of each translated sequence correspond to the A of the initiator codon and the U of the termination codon respectively. In the early mRNAs the number of bases removed by splicing and the extreme nucleotides of each intron are also indicated. The bracket marked *tsA* shows the region of the large T-antigen gene in which most of the *tsA* mutations map. O_R marks the origin of viral DNA replication. Adapted from Tooze (1980).

Figure 1 shows a functional map of SV40 DNA which incorporates most of the information necessary for an understanding of this chapter. The viral genome is a covalently closed, double stranded, circular DNA molecule of 5243 base pairs (bp). The location of particular features on the genome has historically been related to a restriction enzyme cleavage map in which the single *Eco*RI site defined the origin of a map in which the genome was divided into 100 fractional units progressing clockwise. The origin of viral DNA replication is thus at map position 0.67. With the availability of the complete sequence it is clearly sensible to define such features in terms of the nucleotide sequence. Unfortunately, several systems of numbering have arisen. In the belief that Tooze (1980) will be the primary reference source for those intending to use SV40 as a vector, I shall adopt the nucleotide numbering system of Buchman, Burnett and Berg which is used in that volume.

2   The alternative life cycles of SV40

SV40 can enter one of two cycles depending on the type of host cell used. In permissive cells, which for almost all practical purposes are established cell lines, e.g. CV-1, derived from African green monkey kidney, a normal productive infection occurs. The infecting virions are transported to the nucleus where uncoating occurs. The first detectable event is the production of two viral transcripts called the early RNAs. The distinction between early and late events is the classic virological one; early events occur before the onset of viral DNA replication, late events occur afterwards. Early RNA is transcribed in an anti-clockwise direction from the section of the viral genome between nucleotides 5242–5222 and nucleotides 2587/2586, which represent the site of polyadenylation. The primary early transcript can be spliced in two ways. One splice removes 66 nucleotides between nucleotides 4638 and 4571 to generate a mature mRNA of 2.6 kb which encodes a protein called small t-antigen. The other removes 346 nucleotides between positions 4918 and 4571 to generate a 2.3 kb mRNA which encodes large T-antigen.

Small t-antigen is not required for the productive infection of cultured cells. Large T-antigen, however, plays a vital role in this process. This multifunctional protein binds specifically to sequences at or close to the viral origin of DNA replication. This binding is required for the initiation of viral DNA replication and, because of the coincidence of the origin and the promoter for the early transcription unit, T-antigen autoregulates its own synthesis.

Once viral DNA replication has begun, production of the late mRNAs is activated; these are transcribed in a clockwise direction from the region encompassing nucleotides 28 to 2674; this latter nucleotide is the polyadenylation site for all late mRNAs. The synthesis of SV40 late mRNAs is a complex matter. These RNAs can be divided into the two classes shown in Fig. 1. The 16S RNAs, in which a colinear body stretching from nucleotide 1463 to nucleotide 2674 is spliced to a leader segment derived from the sequences immediately on the late side of the origin, code for the major capsid protein VP1. In the 19S mRNAs much smaller segments are removed by splicing and one subclass is not spliced at all; these mRNAs encode the minor capsid proteins VP2 and VP3. In both classes of

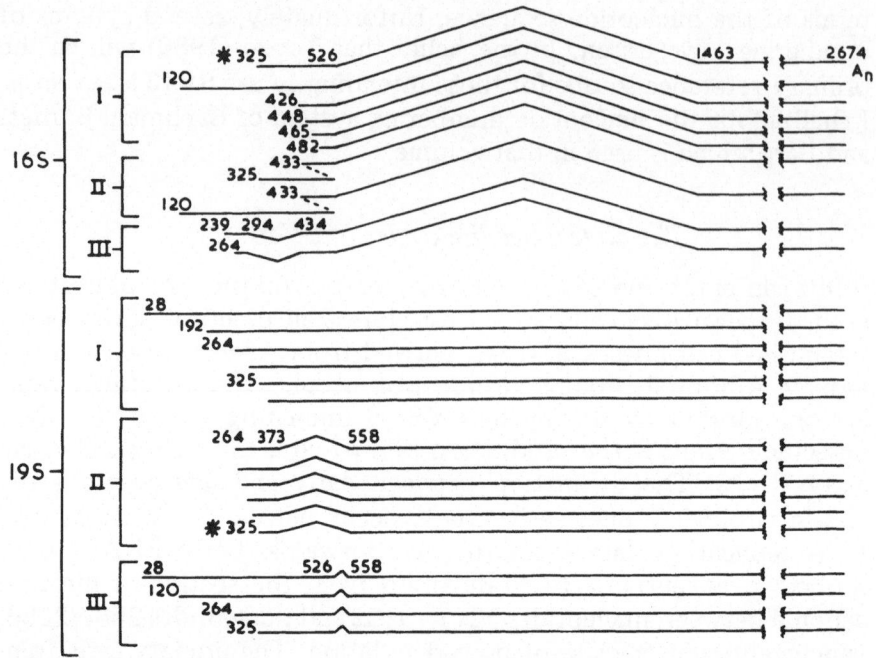

Figure 2   The late mRNAs of Simian virus 40. The mRNAs are divided into the
16S size class, which encodes VP1, and the 19S size class, which encodes VP2
and VP3. All 16S mRNAs contain a splice between nucleotides 526 and 1463;
the members of subclass II contain a duplication of sequences between nucleotides 433 and 526, while the members of subclass II contain an additional splice
between nucleotides 294 and 434. Within the 19S class, subclass I mRNAs are
unspliced, subclass II mRNAs contain a splice between nucleotides 373 and 558
and subclass III mRNAs contain a splice between nucleotides 526 and 558. In
those cases in which the 5′ termini are not numbered present mapping data
identify the species but not its exact terminus. The two species marked *, with
5′ termini at nucleotide 325, are the most abundant members of each class. All
late mRNAs are polyadenylated at nucleotide 2674.

late mRNA a multiplicity of capped 5′ ends has been mapped and
there are several different patterns of splicing. Figure 2 shows a
simplified representation of the 5′ ends and splice junctions of SV40
late mRNAs. Because the location of splice junctions is a paramount
consideration in the design of viral vectors, those planning the use of
SV40 late replacement vectors are strongly advised to consult the
original literature on this point. The situation is further complicated
by the fact that many deletions within the region covered by the late
5′ ends are viable. The late transcription unit, which, interestingly,
does not contain a "TATA" box, is an extremely plastic entity.

Once late transcription has been activated the three capsid proteins
are synthesized, the newly replicated viral DNA is encapsidated and
infectious progeny virions are released with concomitant cell death.

There are fairly strict limits on the size of DNA that can be packaged into a SV40 virion. Molecules as short as 70% of the wild-type length can be encapsidated, but in stocks of such "deleted" genomes duplications of viral sequences which restore the wild-type length are common. DNA molecules longer than the viral genome are packaged extremely poorly. It is thus necessary to design recombinant SV40 genomes so that they are as close as possible to the size of wild-type viral DNA.

Mouse and rat cells are non-permissive for SV40; the viral replicative cycle is blocked at the level of DNA replication. Although the early transcription unit is efficiently expressed the cells lack permissivity factors which are required, together with large T-antigen, for viral DNA replication. The absence of replication means that the late functions are not efficiently expressed and no progeny virions are produced. The majority of the infected cells survive and of these survivors some proportion manifest a heritable series of alterations in their growth properties which defines the transformed state. Such transformed cells can be readily selected by virtue of their ability to overgrow a monolayer of normal cells, to divide in low concentrations of serum and to grow in an anchorage-independent fashion, i.e. to proliferate in semi-solid medium in the absence of attachment to a solid substratum. In transformed cells viral DNA is covalently integrated into the chromosomal DNA of the host cell and the synthesis of large T-antigen from such integrated templates is required for the maintenance of the transformed state. Small t-antigen is not required for transformation *in vitro* or for tumour induction *in vivo*.

It is now clear that the integration of SV40 DNA is frequently accompanied by rearrangements of the viral genome (Botchan *et al.*, 1980; Clayton and Rigby, 1981). Moreover, such rearrangements, and also amplification of both the integrated viral DNA and adjacent cellular sequences, can continue long after the initial transformation event (Hiscott *et al.*, 1980, 1981; Sager *et al.*, 1981; Bender and Brockman, 1981; Clayton and Rigby, 1981). Such alterations in the structure of the transforming genome may well be of functional significance because many transformed cell lines contain viral mRNAs and T-antigens not found in productively-infected permissive cells (McCormick *et al.*, 1980; Rigby *et al.*, 1980; May *et al.*, 1981). The use of SV40 as a vector to integrate exogenous DNA into the chromosomes of the host cell is therefore subject to many uncertainties.

In all infected cells unintegrated SV40 DNA exists as a mini-chromosome in which the viral genome is complexed with the histones of the host cell in a form which closely resembles the nucleosomal structure of cellular chromatin. However, there are distinct differences between the viral chromatin and bulk cellular chromatin and the discovery of the enhancing effect of certain SV40 sequences on the expression of linked cellular genes (Banerji *et al.*, 1981) raises the possibility that there may be viral sequences capable of reorganizing chromatin structure. The use of SV40 vectors to study chromatin organization on inserted genes may therefore be fraught with difficulties of interpretation. The various forms of SV40 chromatin, and their implications for the use of the viral genome as a vector, have been reviewed by Elder *et al.* (1981).

3   Early work with SV40 vectors

Genetical and biochemical analyses have identified a segment of 85 bp, including the origin, as the only *cis*-acting sequences required for the replication of SV40 DNA; all other functions can be provided in *trans*. There exists a class of naturally occurring defective SV40 DNA molecules in which the genome consists of a series of tandemly repeated units; each unit contains some host DNA and the region of the viral genome surrounding the origin. Individual repeats thus provide potential cloning vehicles and were utilized in some of the earliest work (Ganem *et al.*, 1976). However, these segments do not carry a genetically detectable SV40 function and thus the recombinants had to be propagated with a wild-type helper virus. The technical problems associated with this strategy mean that it is no longer used.

All more recent work has employed defective viral vectors in which foreign DNA is inserted into either the early or late transcription unit thus inactivating the expression of viral genes. Insertion into the late unit has been most widely used because in productively-infected permissive cells there is a much higher level of transcription from the late promoter than from the early promoter and thus the detection of the RNA and protein products of the inserted DNA is facilitated. During early studies of the genetic organization of SV40 methods were developed for the growth of absolutely defective viral genomes by complementation with a conditionally defective, temperature-sensitive, helper virus. If a cell is co-infected with a recombinant in which the late region has been replaced by foreign DNA and a *tsA* mutant which is defective for early function at 41°C, then at this non-permissive temperature growth can occur by complementation. The recombinant provides early functions, the helper virus provides late functions and a mixed virus stock results. In cells which receive only the recombinant or only the helper the infection will be abortive. This technique is used not only to propagate stocks of recombinant virions but also to plaque purify individual recombinants from heterogeneous mixtures.

Goff and Berg (1976) and Hamer *et al.* (1977) used this method to clone segments of prokaryotic DNA in late replacement vectors and thus demonstrated the feasibility of the over-all approach. In these, and many other, early experiments the actual cloning was performed in eukaryotic cells. Goff and Berg isolated the larger segment of SV40 DNA from the *Hpa*II site at nucleotide 346 to the *Bam*I site at nucleotide 2533; this vector, which contains all of the early region and almost none of the late region, is called SVGT1 (Fig. 3). The desired segment of λ DNA was also purified by gel electrophoresis and the vector and insert DNAs were joined by homopolymer tailing. The hybrid DNA molecules were then transfected, together with the DNA of the *tsA* mutant helper virus, into permissive monkey cells and mixed plaques, produced by complementation, were picked and used to produce mixed virus stocks. DNA isolated from cells infected with such mixed stocks was assayed for the presence of λ sequences by solution hybridization. This tedious procedure was greatly simplified when Villarreal and Berg (1977) developed an *in situ* technique for screening SV40 plaques by hybridization. This method is analogous to the widely used procedure for the screening of λ plaques (Brammar, this volume). Even with this development, performing the primary cloning in animal cells is difficult, time-consuming and expensive in terms of cell culture media. Now, therefore, the desired recombinant is almost always constructed in *E. coli*. The vector segment of SV40 DNA is first cloned in a plasmid, modified if necessary, and the foreign DNA is then inserted. Finally, the recombinant viral genome is excised from the *E. coli* plasmid vector and transfected into animal cells. One can thus be certain that all events involve the desired recombinant and secondary screening procedures are eliminated.

Hamer *et al.* (1977) used procedures similar to those of Goff and Berg (1976), although the two DNAs were joined via restriction enzyme generated cohesive termini, to introduce into a vector comprising the larger segment of SV40 DNA from the *Hpa*II site to the *Eco*RI site at nucleotide 1782 a segment of prokaryotic DNA including an *E. coli* tRNA gene. They detected in cells infected by the recombinant 19S transcripts containing covalently linked SV40 and *E. coli* sequences but the structures of these RNAs were not documented. Goff and Berg were unable to detect expression of the inserted prokaryotic segment, although Goff (1977) did subsequently show that infected cells synthesize unstable nuclear RNA containing λ sequences.

A particular concern in the work of Goff and Berg was that

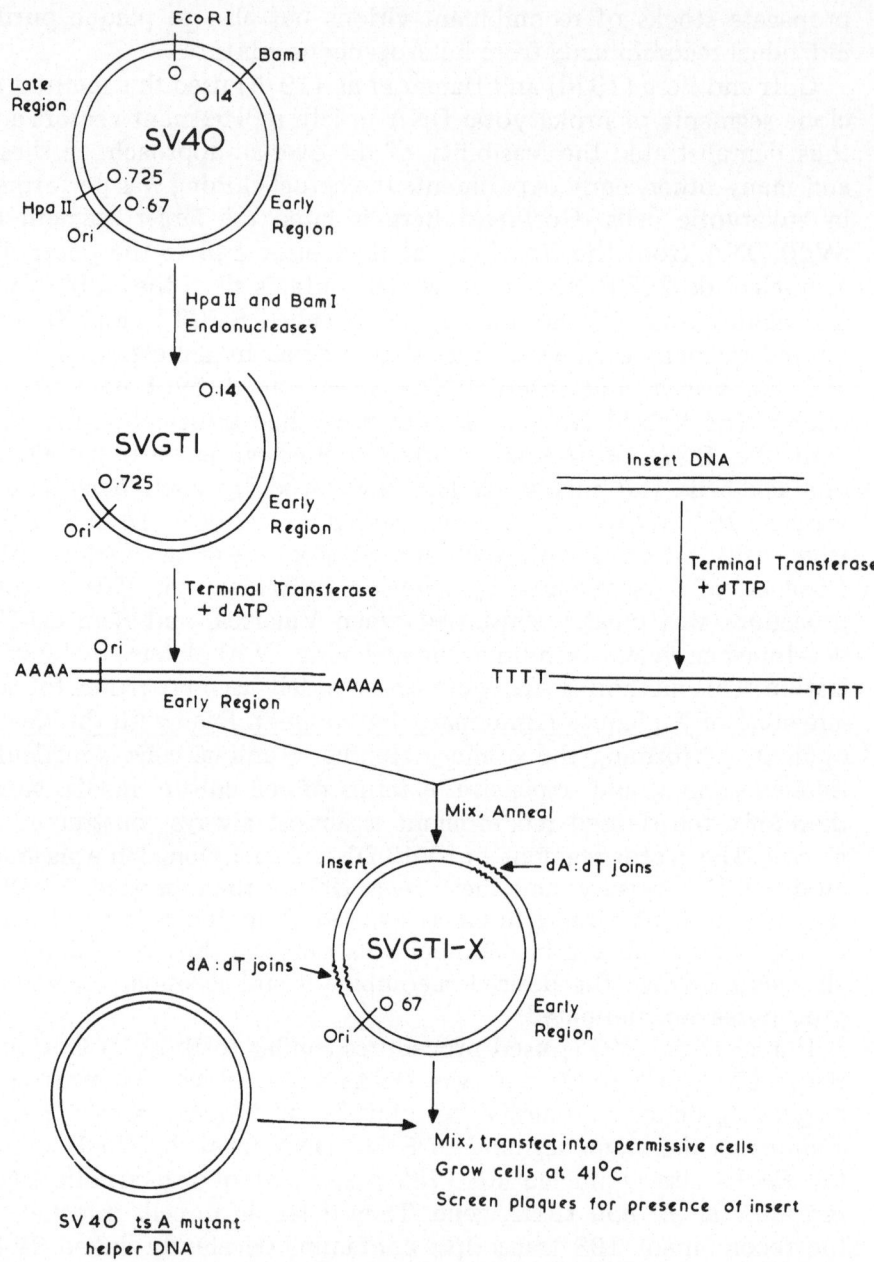

Figure 3   Protocol for cloning foreign DNA in the Simian virus 40 late replacement vector SVGT1 using homopolymer Tailing. Restriction endonuclease cleavage sites and the viral origin of DNA replication are indicated in map units. The single *Eco*RI cleavage site (nucleotide 1782) defines the position 0/1, and the map is divided into 100 fractional units. This protocol is taken from the work of Goff and Berg (1976).

expression of the λ DNA might be prevented by the poly dA.dT tails used for insertion. Goff (1977) eliminated this possibility by showing that he could use homopolymer tailing to reinsert the late genes excised during the preparation of SVGT1 and that the inserted segment was expressed so long as it was in the wild-type orientation. Expression was not affected by which strand carried the polyA segment and the mRNAs synthesized included transcripts of the homopolymer tails. Goff and Berg (1979) used similar procedures to clone the tk gene of *E. coli* and a tRNA^{tyr} gene from *S. cerevisiae*. They could detect heterogeneous nuclear RNA complementary to the inserted prokaryotic sequence, but the *E. coli* tk enzyme was not synthesized. However, the *S. cerevisiae* recombinants did direct the synthesis of a small RNA molecule which had similar electrophoretic properties to yeast tRNA^{tyr} and which hybridized to a cloned yeast tRNA^{tyr} gene. tRNA is normally transcribed by RNA polymerase III; the precise mechanism by which this small RNA was produced was not explored, although it is interesting to note that its synthesis was independent of the orientation of the insert, suggesting that a promoter on the yeast DNA was active.

In all of these experiments the investigators lacked one vital piece of information, that SV40 late mRNAs are spliced. Moreover, splicing appears to be an obligatory step in the biosynthesis of most of the stable, late, cytoplasmic RNAs (Lai and Khoury, 1979; Gruss *et al.*, 1979). In SVGT1 the late splice junctions are deleted and the lack of expression is, with hindsight, easily understood. The location of the late splice junctions was rapidly established and with this information to hand it was then relatively simple to design vectors in which the inserted DNA is positioned so that the SV40 promoters and splicing signals are intact.

4   Late replacement vectors

Mulligan *et al.* (1979) were the first to exploit knowledge of the arrangement of the late splicing signals by constructing a recombinant in which the coding sequence for VP1 was precisely replaced by a cDNA copy of the mRNA for rabbit β-globin. This SVGT5-based recombinant (see Fig. 8) was propagated in permissive monkey cells by complementation with a *tsA* mutant helper. The recombinants direct the synthesis of the expected hybrid 16S mRNA which contains the SV40 late leader and splice junctions but a 3′ exon comprised mainly of the inserted globin sequences. The recombinant was constructed so that it contains only the coding region of the globin cDNA; polyadenylation is specified by viral sequences. The hybrid mRNA is translated to give a protein indistinguishable from authentic rabbit β-globin. Although large amounts of globin are synthesized, comparable to the amounts of VP1 produced in a wild-type infection, the protein is highly unstable, presumably because of the absence of the α subunit and of the haem prosthetic group. Analogous experiments leading to similar conclusions were also performed by Hamer *et al.* (1979).

Subsequently a number of other recombinant viral genomes of similar design have been constructed and analysed. Hamer and Leder (1979a) and Hamer *et al.* (1980) used genomic globin DNA rather than cDNA and again observed the production of authentic globin mRNA and protein. Gruss and Khoury (1981) inserted a segment of rat genomic DNA coding for preproinsulin and showed that the precursor protein is synthesized and processed to proinsulin with concomitant secretion into the culture medium. Not surprisingly, the monkey cells are incapable of processing the proinsulin to the mature hormone. Gruss *et al.* (1981c) have also constructed recombinants which express the p21 transforming protein of the Harvey murine sarcoma virus (HaMSV).

In experiments involving genomic DNA rather than cDNA there are two possible mechanisms for transcription of the inserted segment. mRNA synthesis could initiate at the SV40 late promoter, with expression of the foreign DNA being due to read-through transcription or, if the genomic segment contains sufficient DNA upstream of the cap site, initiation could occur at the promoter carried on the insert. Hamer *et al.* (1980) have shown that in recombinants carrying the mouse α-globin gene in SV40 late replacement vectors some of the globin-specific transcripts have the same 5' end as authentic globin mRNA and are thus presumably initiated at the globin promoter. However, globin-specific transcripts initiating at the SV40 late promoter were also detected. In those cases where the primary object is to study the mechanism of transcription of the inserted gene it will be desirable to construct vectors from which the SV40 late promoter has been removed. Similar problems are associated with the analysis of the 3' ends of transcripts from such recombinants. The commonly used late replacement vectors retain the AATAAA sequence which acts as a signal for the polyadenylation of SV40 late transcripts. If the inserted genomic DNA also contains an AATAAA sequence there are two possible sites for polyadenylation and transcripts polyadenylated at both the insert and viral sites have been observed in cells infected with SV40 mouse α-globin and SV40 rat preproinsulin recombinants (Hamer *et al.*, 1980; Gruss and Khoury, 1981). The use of either of two possible polyadenylation sites is not unexpected. In an analysis of the precise sequence requirements for the polyadenylation of SV40 late mRNAs,

Fitzgerald and Shenk (1981) constructed viral genomes in which the putative polyadenylation signal is tandemly duplicated and showed that both copies are functional.

SV40 late replacement vectors have been widely used to analyse the role of splicing in the biogenesis of stable, cytoplasmic mRNA. Deletion mutants of SV40 in which the late splice junctions are absent fail to produce stable, cytoplasmic mRNA although nuclear transcripts can be detected (Lai and Khoury, 1979; Gruss *et al.*, 1979). Gruss *et al.* (1979) cloned a cDNA copy of VP1 mRNA into a late replacement vector lacking the splice junctions, thus generating an intronless late transcription unit, and were unable to detect any synthesis of VP1. If an intron derived from the mouse β-globin gene is inserted into this intronless mutant then the synthesis of VP1 is restored (Gruss and Khoury, 1980). This report supports the idea that splicing is necessary for the production of stable, cytoplasmic mRNA and indicates that heterologous introns function efficiently. Hamer and Leder (1979b, c) have also investigated the role of splicing. They cloned segments of the mouse β^{maj}-globin gene, either containing or not containing an intron, into vectors which either did or did not contain the SV40 late splice junctions and obtained a series of recombinants representing all possible orientations of the vector and insert splice junctions (Fig. 4). This work demonstrated that cytoplasmic mRNA is produced only if the recombinant contains at least one intact splice junction.

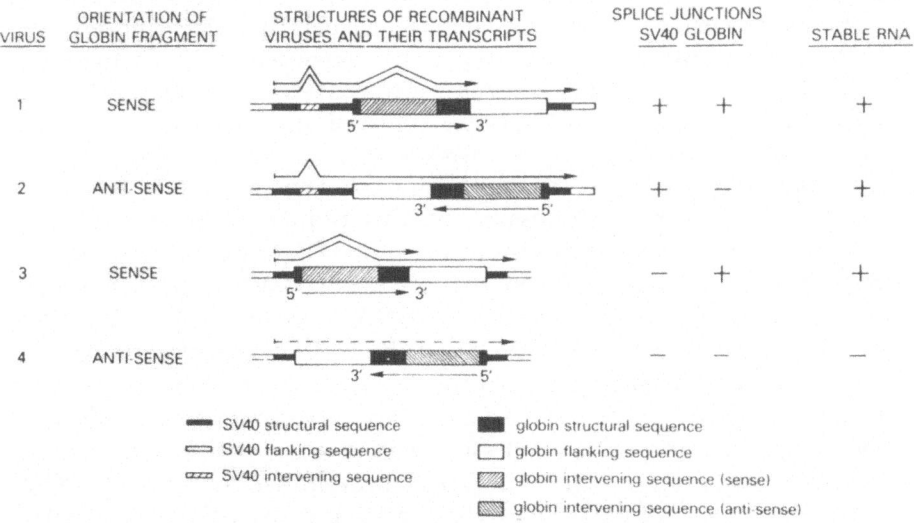

Figure 4 The use of Simian virus 40-globin recombinants to study the role of splicing in stable mRNA synthesis. Reproduced, with permission, from Hamer and Leder (1979c).

Such experiments suggest an obligatory role for splicing in mRNA synthesis. However, some eukaryotic genes, e.g. those for histones and interferons, do not contain introns and splicing of the relevant transcripts has not been detected. Yet these genes are expressed, in the case of histone genes at high levels. Moreover, Treisman *et al.* (1981) have shown that a recombinant plasmid containing an early region of polyoma virus in which the splice junction used in the synthesis of the mRNA for middle T-antigen has been replaced by the corresponding section from a cloned cDNA copy of the mRNA can transform rat cells, thus proving that middle T-antigen is the transforming protein of polyoma. Endonuclease S1 mapping of the viral transcripts in such transformed cells shows that the middle T-antigen mRNA is not spliced. Similarly, Gruss *et al.* (1981b) have used the SV40 rat preproinsulin recombinant discussed above to construct a derivative lacking the preproinsulin gene intron. This recombinant efficiently directs the synthesis of rat preproinsulin. The haemagglutinin gene of influenza virus has also been expressed from vectors lacking splicing signals (Gething and Sambrook, 1981; Sveda and Lai, 1981).

The requirement for splicing is thus unclear. Some genes are naturally expressed in the absence of splicing. Those genes which are normally spliced may, or may not, be efficiently expressed when the introns are removed. When using SV40 vectors to express inserted DNA it would seem prudent to always perform the initial construction in such a way as to retain intact splicing signals. If expression is obtained the effect of deleting such signals can then be explored.

A further factor which might be expected to influence the efficiency with which an inserted gene is expressed is the positioning of the AUG codon which initiates translation. It has generally been held that in potentially polycistronic eukaryotic mRNAs only the initiator codon closest to the 5' end of the mRNA is used (see Kozak, 1980, for a discussion of this point). However, in their experiments on the expression of rabbit β-globin cDNA in late replacement vectors Mulligan *et al.* (1980) also used the vector SVGT7 (see Fig. 8) in which the cDNA is positioned so that it is transcribed into 19S mRNAs. These hybrid transcripts contain the AUG codons normally used to initiate translation of VP2 and VP3 upstream of the globin AUG. However, in cells infected with such recombinants authentic β-globin was again synthesized; hybrid proteins containing both viral capsid protein sequences and globin sequences were not detected, indicating that translation was initiated at an internal AUG. Similarly, Subramani *et al.* (1981), studying the expression of a cDNA copy of mouse dihydrofolate reductase mRNA cloned in SVGT7 observed the synthesis of authentic dihydrofolate reductase protein from a mRNA containing upstream viral initiator codons. Mulligan and Berg (1981b) have further shown that the *E. coli gpt* gene can be translated from hybrid mRNAs in which the coding sequence begins approximately 800 nucleotides from the 5' end of the mRNA and the effective initiation codon is preceeded by five AUG codons. While these results cast doubt on the view that only the 5'-proximal AUG can be used to initiate translation, recombinants specifying hybrid mRNAs will provide a powerful assay system for the analysis of the precise signals which specify the initiation of eukaryotic translation.

5   Early replacement vectors

SV40 early replacement vectors have been much less widely used than late replacement vectors, mainly because the early transcription unit is expressed at a much lower level than the late unit in productively-infected permissive cells. However, the use of early replacement vectors has acquired a new dimension with the development by Gluzman (1981) of COS cells. CV-1 cells are a permissive line of monkey kidney cells and infection with wild-type SV40 leads to a productive cycle and cell death. These cells can, however, be transformed by UV-inactivated SV40; transformation is induced by mutant viral genomes in which the replication function of large T-antigen is defective. Gluzman cloned the early region of SV40 DNA in a plasmid vector and then, rather than mutating the T-antigen gene, he mutated the origin *in vitro*. CV-1 cells were then transformed with this origin-defective early region which can not replicate even in the presence of wild-type T-antigen. The resulting COS cells express wild-type large T-antigen and contain the permissivity factors required for SV40 DNA replication. Infection of COS cells by early deletion mutants, which can not make T-antigen, leads to a normal lytic infection because the mutant genomes are replicated under the influence of the wild-type large T-antigen produced from the resident viral early region (which can not itself replicate or be rescued) and then express late functions. In this way it is possible to produce pure virus stocks of early deletion mutants and of recombinants in which the early region has been replaced by foreign DNA. The COS cell system obviates the need for a temperature-sensitive helper virus and the ability to produce pure stocks of recombinant virions has obvious technical advantages. One potential problem with this system is that permissive cells containing large amounts of T-antigen are somewhat unhealthy and COS cells are thus more difficult to grow than normal CV-1 cells. Moreover, it is possible that there will be recombination between the early region sequences that are retained in most vectors to provide transcriptional control signals and the resident early region. Such difficulties notwithstanding, it is clear that this system will be of great value, not only for the propagation of recombinant virions but also as a host for episomal vectors which depend on the SV40 origin for their replication (see Section IV.C.2).

Early replacement vectors, like their late counterparts, must be designed with proper regard for the location of viral transcriptional control and splicing signals. The most common tactic is to join the 5' end (in the transcriptional sense) of the foreign DNA to the *Hind*III site at nucleotide 5171. This site lies between the SV40 early cap sites (nucleotides 5242–5222) and the AUG codon which initiates translation of the T-antigens (nucleotides 5164/5162). The inserted DNA will then be transcribed by read-through from the SV40 early promoter. Polyadenylation can be specified either by a signal on the inserted DNA or by retaining in the vector the polyadenylation signal used in the production of SV40 early mRNAs. In one respect early replacement vectors are more complex than late replacement vectors. This is because the early region introns are within the sequences coding for the T-antigens and are thus removed during vector construction. If the inserted segment is to be cDNA an intron must be restored. The common solution to this problem is to take a small segment of SV40 DNA containing an early region intron and insert it into the vector downstream of the foreign DNA (Berg, 1981; Gething and Sambrook, 1981).

6   Novel uses of SV40 vectors

The studies discussed so far have been designed to use SV40 vectors in order to amplify eukaryotic genes so that the mechanisms of transcription and of post-transcriptional processing can be more easily analysed. The available technology has, however, strong implications for two quite different areas of research. Firstly, the ability to express a single gene at a high level in (theoretically) any cell type should allow the dissection of multiple factor pathogenesis. Secondly, the expression from viral vectors of appropriate genes will facilitate studies of the biosynthesis and assembly of complex subcellular structures and organelles. Both of these areas are exemplified by recent studies in which SV40 vectors have been used to express the haemagglutinin (HA) of influenza virus (Gething and Sambrook, 1981; Sveda and Lai, 1981) and the surface antigen of human hepatitis B virus (Moriarty *et al.*, 1981). Sveda and Lai used the standard late replacement vector approach to construct a recombinant in which a full length cDNA copy of vRNA segment 4 (which encodes haemagglutinin) from an influenza A strain virus was inserted between the *Hae*II site at nucleotide 832 and the *Bam*I site. The recombinant was propagated as a mixed stock with a *tsA* mutant helper virus. Cells infected with the recombinant synthesize the precursor molecule HA_0, which is glycosylated, but cleavage to the functional HA_1 and HA_2 subunits was not observed. Even so the HA produced is capable of agglutinating guinea pig erythrocytes and was shown by indirect immunofluorescence to be located in the plasma membrane.

Gething and Sambrook adopted a somewhat different strategy. They manipulated a cloned cDNA copy of an influenza A strain HA gene so that the initiator AUG was positioned immediately adjacent to the restriction enzyme site to be used for cloning. This segment was then inserted into a late replacement vector, between the *Hpa*II and *Bam*I sites, so that the HA gene was correctly orientated for transcription from the SV40 late promoter. In a parallel series of experiments they inserted the cloned HA gene into an early replacement vector immediately downstream of the SV40 early promoter. Immediately after the HA gene they inserted the intron normally used during the synthesis of the mRNA for small t-antigen and the SV40 early polyadenylation signal, thus ensuring that the transcript of the HA gene could be appropriately processed. In both cases all of the constructions were performed in *E. coli* and were verified before the recombinants were excised from the bacterial plasmid vector and transfected into eukaryotic cells. This contrasts with the approach of Sveda and Lai who cloned the HA segment directly into SV40; not surprisingly, they observed that their recombinant virus stock was heterogeneous and that only 1 in 9 of their recombinant clones directed HA synthesis. Gething and Sambrook propagated their late replacement vector by complementation with an early region deletion mutant of SV40 and their early replacement vector in COS cells, thus generating a pure recombinant virus stock.

Cells infected with the mixed virus stock containing the late vector recombinant synthesize HA, detected by radioimmunoassay, at a high level, $80\,\mu g$ of protein/90 mm dish or 6×10^8 molecules/cell. Again the HA produced is glycosylated but not cleaved, and indirect immunofluorescence indicates that it is located at the cell surface and within intracellular organelles, which are probably Golgi bodies. The HA produced was functional in agglutinating erythrocytes but was not shed into the medium, as it is in influenza virus-infected cells. COS cells infected with the early vector recombinant synthesize HA with the same properties, but the amount is much lower. Gething and Sambrook present data which suggest that only a proportion of the infected cells contain enough T-antigen to support the replication of the recombinant; moreover, one would expect a lower level of synthesis from the early promoter.

The ramifications of this type of system are considerable. Such recombinants allow the effects of particular viral components on host cells to be studied in isolation and thus it should be possible to deduce the role of each component in the infectious process and in viral pathogenesis. The results of Sveda and Lai and of Gething and Sambrook show that insertion of HA into the membrane requires no other viral components. Similar recombinants can be constructed from other influenza vRNA segments and thus the interactions of the viral polypeptides can be analysed. Coinfection of cells with recombinants expressing HA and the matrix protein will reveal whether interactions between these two proteins are required for HA to be budded into the medium. The cloned HA gene can also be subjected to site-directed mutagenesis *in vitro* and such altered genes similarly expressed in cultured cells. This approach will facilitate analysis of the sequences required for correct membrane insertion and also of the regions involved in antigenicity. Finally, it may be possible to fuse sequences coding for other proteins into the HA gene and thus ensure the membrane insertion and/or secretion of proteins which never normally interact with membranes.

A SV40 late replacement vector has also been used to express the surface antigen of human hepatitis B virus (HBV) (Moriarty *et al.*, 1981). Although the recombinant was constructed by cloning in *E. coli* it retains plasmid vector sequences; following propagation with a *tsA* mutant helper virus only 5% of the virus stock was the desired recombinant. 75% was helper virus while the remainder comprised a heterogeneous collection of recombinant DNA molecules containing HBV sequences. Despite this somewhat inelegant procedure, cells infected with the recombinant virus mixture synthesize surface antigen, 40% of which is found in the medium. The secreted antigen has the physical properties of the 22 nm particles found in the serum of human hepatitis carriers. The morphology of these particles was confirmed by electron microscopy, and gel electrophoresis showed that they have the same polypeptide composition as authentic 22 nm particles. The yield of surface antigen is rather low but it could surely be increased by more precise construction of the recombinants. However, the potential of this type of experiment is readily apparent. Individual HBV components can be produced in cultured cells in which they are glycosylated and properly assembled into particles, processes which do not occur when the surface antigen is expressed in *E. coli*. The study of the biology of HBV will thus be greatly facilitated. Moreover, such eukaryotic systems might have considerable advantages for vaccine production.

B   Human adenoviruses

1   Introduction

Human adenoviruses have been widely canvassed as potential vectors. "Natural" experiments in genetic engineering, performed many years ago during the production of adenovirus vaccine stocks in rhesus monkey cells which were contaminated with SV40, showed clearly the potential of adenoviruses for the expression of exogenous genetic information (Sambrook and Grodzicker, 1980). The intentional use of adenoviral DNA as a vector is, however, technically rather more difficult than the use of SV40 DNA and so adenovirus vector systems remain at an early stage of development.

2   The life cycles and molecular biology of adenoviruses

The molecular biology of human adenoviruses, particularly of serotypes 2, 5 and 12, has been intensively studied over the last decade and has been comprehensively reviewed in Tooze (1980). Adenovirions contain a linear, double stranded DNA genome which in the case of serotype 2 (Ad2) is 35 kb in length. Of particular relevance to the possible use of this genome as a vector is the fact that the 5′ termini of adenoviral DNA strands are covalently linked to a protein, which in its mature form has a molecular weight of 55 000. Studies of adenovirus DNA replication *in vitro* indicate that this protein plays a crucial role and it is widely believed to perform a priming function. Moreover, DNA which retains the terminal protein is more infectious in transfection experiments than DNA from which the protein has been removed by proteolytic digestion. In order to clone the terminal fragments of an adenoviral genome it is necessary to remove this protein, which would otherwise block ligation to the vector, without removing adjacent DNA sequences.

In productively-infected human cells the expression of the adenoviral genome is temporally regulated, in a manner analogous to that previously discussed for SV40. However, the situation is distinctly more complex; adenoviruses contain not one but five early transcription units and also several late transcription units, one of which predominates (Fig. 5). Within a given transcription unit the primary, nuclear transcript can be spliced in a number of ways; a further complication is introduced by the fact that a given transcription unit can contain multiple polyadenylation sites. Thus, when RNA polymerase II initiates transcription at the major late promoter at map position 16.6 transcription continues to the right end of the genome, each polymerase molecule traversing the complete unit once.

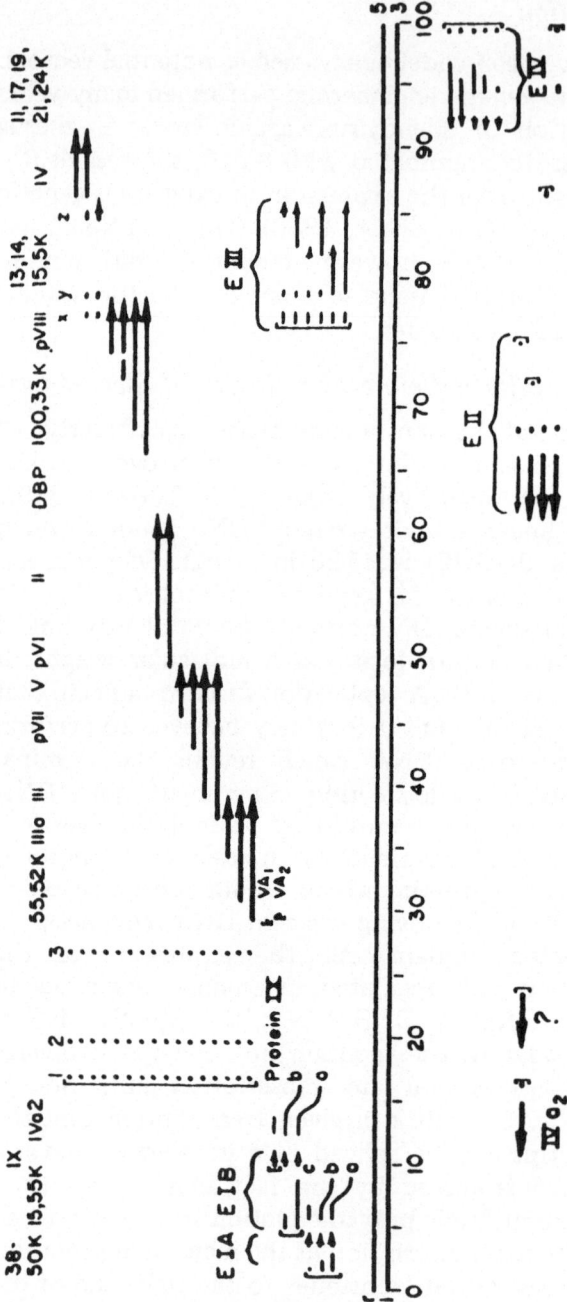

Figure 5 Transcription map of human adenovirus type 2. Early mRNAs, which are organized into five transcription units, EIA to EIV, are indicated by thin arrows. Within each early transcription unit individual mRNAs are generated by multiple splicing events. Late mRNAs are indicated by thick arrows. All the late mRNAs transcribed in the rightward direction initiate at the major late promoter at map position 16.6 and contain the sequences, designated 1, 2 and 3, of the tripartite leader. x, y, and z indicate additional leader sequences found in some late mRNAs. The mRNAs derived from the major late transcription unit are organized into five 3' co-terminal families. VA₁ and VA₂ are RNA polymerase III transcripts. At the top of the map are shown the approximate locations of protein coding sequences. This map was originally compiled by T. Broker and L. Chow. This figure is reprinted, with permission, from *Nature, Lond.* **287**, (1980) 492. Copyright © 1980 Macmillan Journals Ltd.

There are five potential polyadenylation sites only one of which is used in a given primary transcript. Shortly after the polymerase has passed such a signal an endonucleolytic cleavage occurs which allows polyadenylation and the complex series of splicing reactions then ensues. The rules which determine the utilization of the various polyadenylation sites are unclear. Within each family of 3' co-terminal RNAs alternative splicing pathways generate a multiplicity of mature mRNAs.

The work with SV40 discussed above has clearly demonstrated that in order to utilize an animal virus genome effectively as a vector the pathways of viral mRNA biogenesis must be clearly understood and the foreign DNA must be inserted so that it is correctly orientated with respect to viral promoters, polyadenylation sites and splicing signals. Inspection of Fig. 5 reveals that in the case of adenoviruses this is not so simply achieved and for this reason only restricted use has been made of adenoviral vectors.

3   Manipulation of the adenoviral genome

It is technically more difficult to manipulate a large, linear genome than the small circular genome of SV40. Many of the routine techniques of recombinant DNA technology presuppose a circular DNA molecule with one or a few sites for the restriction endonucleases to be used during manipulation. One major breakthrough in the use of adenoviral DNA was achieved by Jones and Shenk (1978, 1979) who showed that it is possible to apply to this large, linear genome the techniques so effectively used for the construction *in vitro* of deletion mutants of SV40. The construction of bacteriophage λ vectors depended upon the removal of particular restriction enzyme sites so as to generate a vector containing only one or two sites for the enzyme to be used in cloning (Brammar, this volume). In the case of λ this could be achieved in a variety of ways: (i) by genetically selecting *in vivo* for pre-existing variants that had lost particular sites; (ii) by utilizing well-characterized deletion and substitution derivatives of λ in which sites had been lost or other, more useful, sites introduced and (iii) by exploiting recombinants between λ and related phages such as ∅80 in which the presence of the related DNA led to the loss or acquisition of the required sites. Strategy (i) can not be directly applied to an animal virus but Jones and Shenk successfully adapted it to adenovirus according to the procedure diagrammed in Fig. 6a. If the genome contains three sites for an enzyme, then following cleavage by that enzyme subsequent religation will regenerate an intact genome at low frequency because the joining of three fragments in the correct orientation is required. If, however, the population

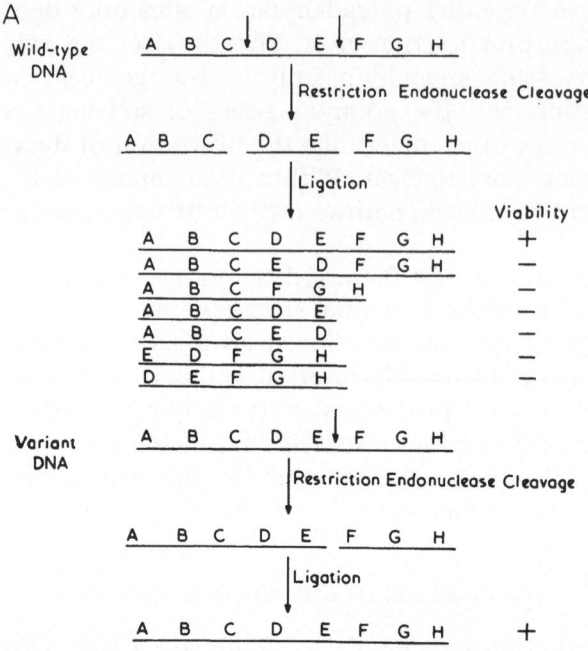

Figure 6   A, Procedure for the isolation for adenovirus mutants lacking restriction endonuclease cleavage sites. This diagram illustrates the procedure utilized by Jones and Shenk (1978, 1979). Only the simplest ligation products derived from wild-type DNA are shown. Other products, containing multiple copies of the (D, E) fragment, are also possible. B, Procedure for the construction *in vitro* of deletion mutants of adenovirus. This diagram illustrates the procedures developed by Stow (1981). The DNA molecules are not drawn to scale.

contains variants lacking one of the sites then the reconstruction of such genomes requires the joining of only two fragments and is therefore much more efficient. By repeated application of this rationale it was possible to construct derivatives of Ad5 containing only a single site for *Xba*I and then to introduce deletion and substitution mutations in the vicinity of this site. In this way it should be possible to generate a wide variety of variant adenoviral genomes in which restriction sites are appropriately positioned with respect to the viral transcription units. Strategy (ii) given above has not thus far been used; the adenovirus analogue of strategy (iii), which exploits interserotypic recombinants, was used in the work of Thummel *et al.* (1981) discussed below.

A second development of major importance was provided by the work of Stow (1981). He showed, using the strategy diagrammed in Fig. 6b, that it is possible to clone the left-terminal fragment of Ad2 DNA, mutate it *in vitro* and then reintroduce the mutagenized fragment into a complete viral genome. The left-terminal *Hpa*I

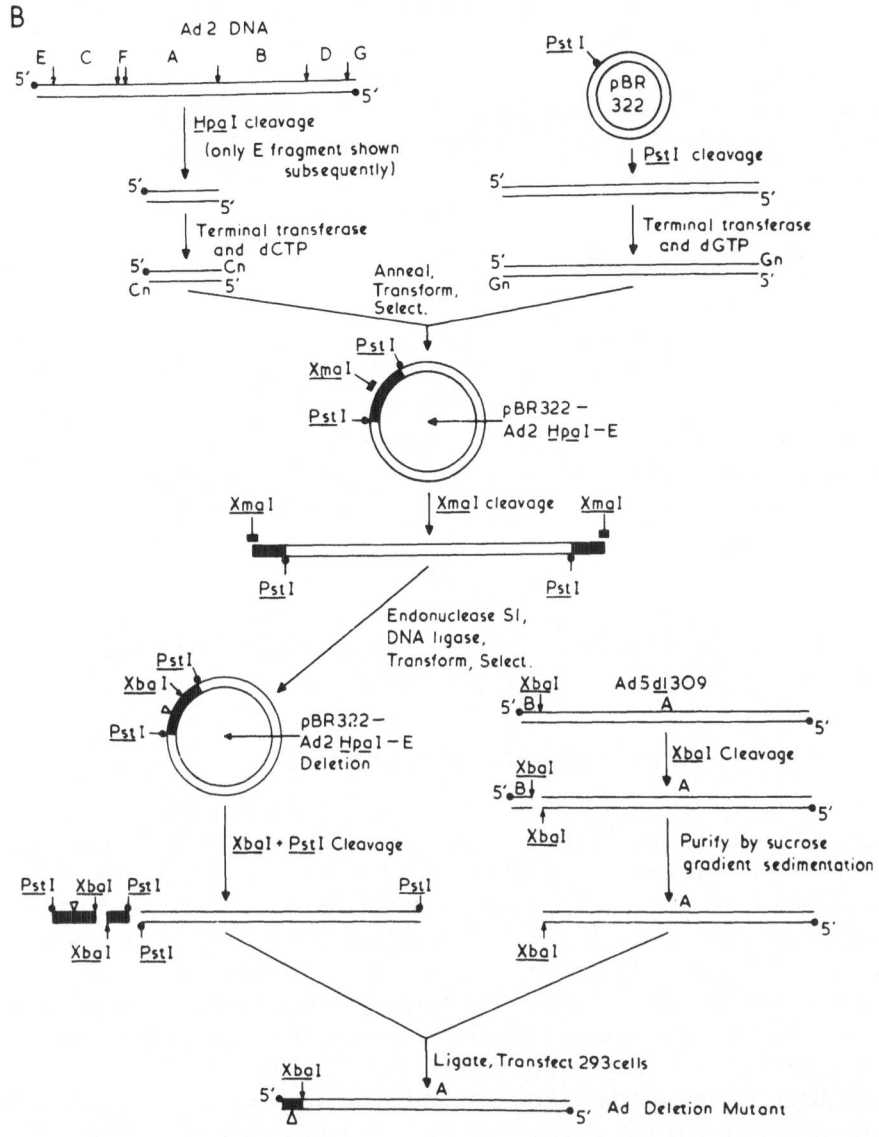

fragment of Ad2 DNA (0 to 4.5 map units), which has a sequence almost identical to that of the corresponding fragment of Ad5, was cloned in a plasmid vector in such a way that no sequences were lost from the left molecular end. Small deletions were then introduced around the *Xma*I site at map position 2.8 and the cloned, mutagenized fragment was cleaved with *Xba*I, which cuts at map position 3.8, and with *Pst*I to separate viral and plasmid sequences. Stow then exploited the Ad5 mutant *dl*309, produced by Jones and Shenk (1979), which retains only one *Xba*I site at map position 3.8. The larger of the two

*Xba*I fragments of *dl*309 was ligated to the mutagenized left-terminal Ad2 fragment and the ligation mixture was transfected into 293 cells (see Section III.B.6). This procedure regenerated intact Ad5 genomes and infectious virions. From further experiments it transpires that it is not necessary to remove the plasmid sequences. If the left-terminal *Xba*I fragment, still carrying vector sequences joined to the left molecular end, is ligated to the larger *Xba*I fragment of *dl*309 and the mixture transfected, the genomes of the viruses which result have exactly the normal left end. Somehow, during replication or packaging, the plasmid DNA is cleanly excised and the attachment site for the terminal protein is regenerated. By applying this methodology it should be possible to construct recombinant adenovirus transcription units entirely in *E. coli*, which will allow the constructions to be performed much more rapidly and precisely. The recombinant transcription units, containing the foreign DNA, can then be reconstructed into adenoviral genomes.

Taken together, these two technical developments should allow adenoviral DNA to be manipulated with great facility and the construction of a wide variety of adenovirus vectors, in which the inserted segment is placed under the control of either early or late promoters, can be confidently expected.

4   Natural adeno-SV40 hybrid viruses

Thus far the use of purpose-built adenovirus vectors has not been reported. There has, however, been an extensive series of "natural" experiments on the use of adenovirus vectors. During the late 1950s adenovirus vaccines were prepared by adapting the virus to growth in rhesus monkey cells. We now know that the infection of monkey cells by human adenoviruses is abortive because of a failure to properly splice late mRNAs. The rhesus monkey kidney cell cultures used were contaminated with SV40 which provides a helper effect, overcoming the RNA processing defect. During the extensive passaging of the vaccine stocks recombination occurred between the infecting adenoviral DNA and the contaminating SV40 DNA to generate a series of hybrid viruses in which segments of the SV40 genome have been covalently integrated into adenoviral DNA. These hybrids, which have been reviewed in Tooze (1980), can be either non-defective, in which cases the SV40 DNA has replaced a non-essential segment of the adenoviral genome, or defective, in which cases essential genes have been replaced and the hybrids must be propagated with a helper virus. These hybrid viruses have been extensively characterized and serve both to begin to define essential and non-essential regions of the adenoviral genome and to act as models for the use of adenoviruses as vectors (Sambrook and Grodzicker, 1980).

5   Constructed adeno-SV40 hybrid viruses

Thummel *et al.* (1981) have reported a series of experiments in which they constructed *in vitro* novel hybrids between Ad2 and SV40. This work did not require purpose-built vectors because there exists a powerful genetic selection for the desired recombinants. They cloned in a plasmid vector a segment of SV40 DNA containing the entire large T-antigen coding region but lacking the early promoter; this construction was performed so that the SV40 DNA insert is bounded by *Bam*I sites. As vector they used the DNA of two adenovirus recombinants, derived from interserotypic crosses between Ad2 and Ad5; these genomes contain only two *Bam*I sites rather than the three present in each parental genome. The vector DNA was cleaved with *Bam*I and ligated to the *Bam*I fragment containing the T-antigen coding sequence; the ligation mixture was then transfected into human cells in the presence of helper virus DNA. The resulting virus stocks were then plaqued on monkey cells. For the reasons described above growth in monkey cells can occur only if they are infected by recombinants which have acquired the SV40 T-antigen gene in such a way that it is expressed under the control of either early or late adenovirus promoters.

Solnick (1981) has used a slightly different strategy to construct a novel hybrid virus expressing SV40 large T-antigen under the control of the Ad2 major late promoter. A segment of SV40 DNA containing the entire T-antigen coding sequence but lacking the early promoter was cloned into a bacterial plasmid carrying a short segment of Ad2 DNA including the major late promoter. The hybrid transcription unit was then excised and joined to the largest *Bam*I fragment of Ad5 DNA (0 to 75.9 map units) and to the right terminal *Bam*I fragment of the non-defective hybrid virus Ad2$^+$ND1 (85.5 to 100 map units). The *Bam*I site at 85.5 in Ad2$^+$ND1 is in fact the *Bam*I site carried in the SV40 insertion. The ligated DNA was transfected, together with intact Ad5 DNA as helper, into human cells and the virus mixture produced was then passaged on CV-1 cells to select for the desired recombinants. Cells infected with the recombinant synthesize large T-antigen from a hybrid transcription unit in which the promoter derives from Ad2 and the splicing and polyadenylation signals derive from SV40.

6   General considerations in the use of adenovirus vectors

In the majority of cases the insertion of foreign DNA into the adeno-viral genome will render the recombinant defective and thus it must be propagated with a helper. This helper could be a wild-type or *ts* mutant virus, but it seems likely that in many cases the helper function will be provided by adenoviral genes integrated into the chromosomes of the host cell. Jones and Shenk (1978; 1979) have already demonstrated the utility of such systems. They constructed a series of deletion and substitution mutants at the *Xba*I site at position 3.8 on the Ad5 map. These mutations inactivate the early transcription unit E1 and the resultant viruses are thus incapable of growth in HeLa cells. However, there exists a line of cells, called 293 cells, which resulted from the transformation of primary human embryonic kidney cells by the left-terminal fragment of Ad5 DNA (Graham *et al.*, 1977). The E1 region mutants grow well in 293 cells because their defects are complemented by expression from the integrated sequences. The use of tk co-transformation (Wilkie, this series, in preparation), or of co-transformation involving other selectable markers (see Section V), should allow the construction of cell lines in which other adenoviral genes are expressed from chromo-somally integrated viral sequences. The first report of such con-structed cell lines expressing adenoviral gene products has already appeared (Grodzicker and Klessig, 1980).

An increasing use of adenovirus vectors is therefore to be expected and this system has considerable advantages over SV40. While the highly complex transcription programme of adenoviruses poses problems in the proper construction of recombinants, it can also be effectively exploited. It should be possible to construct a series of recombinants in which a given gene is expressed under the control of various early and late promoters and thus the level of expression of that gene can be efficiently controlled. Further technical advantages of this system are that adenoviruses can be grown in large-scale spinner cultures of HeLa cells and that adenoviruses, unlike SV40, inhibit host cell protein synthesis. The purification of large quantities of the product of the cloned gene is thus greatly facilitated.

C   Retroviruses

1   Introduction

C-type retroviruses have single stranded RNA genomes but their replication cycle involves an obligatory double stranded DNA

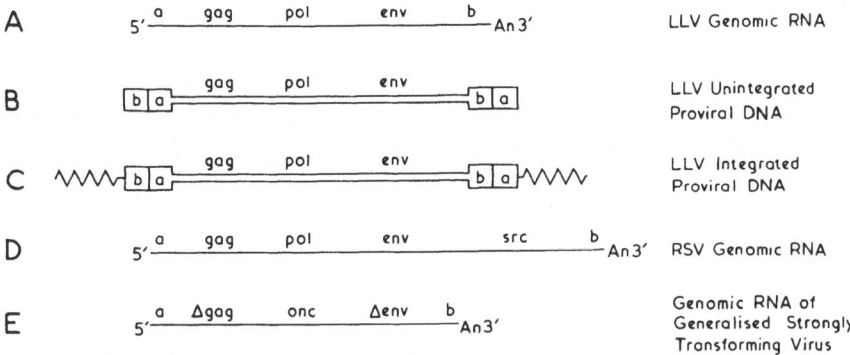

Figure 7   Structures of C-type retroviral genomes and proviruses. The *gag* gene encodes proteins of the virion core, the *pol* gene encodes reverse transcriptase and the *env* gene encodes glycoproteins of the viral envelope. *src* is the transforming gene, encoding pp60^{src}, of RSV. *onc* denotes the transforming gene of a generalized strongly transforming retrovirus; during the generation of such a virus either the gag gene, or the env gene, or both, generally suffer deletions. a and b indicate the sequences at the 5' and 3' termini of genomic RNA which are duplicated during reverse transcription to give rise to the LTRs. ‸⋀⋀⋀ indicates cellular DNA. LLV stands for lymphoid leukosis virus.

intermediate. Both unintegrated and integrated proviral DNA forms of many retroviruses have now been cloned and thus these genomes can be readily manipulated. The biology and molecular biology of retroviruses has been an area of immense progress in recent years and our present knowledge of these viruses will be comprehensively reviewed in Weiss *et al.* (1982).

The basic structure of a retroviral genome is shown in Fig. 7a. Such a leukosis virus, be it of avian or mammalian origin, contains three genes: *gag*, which specifies proteins of the virion core; *pol*, which specifies the reverse transcriptase and *env*, which specifies the glycoproteins of the viral envelope. A virus of this type contains all of the information required for replication but it will not transform cultured fibroblasts. The first step in the replication cycle is the synthesis, by reverse transcriptase, of a linear, double stranded DNA provirus. During this process sequences from the termini of the RNA genome are duplicated to give the structure shown in Fig. 7b. The duplicated structures produced during reverse transcription are known as long terminal repeats, or LTRs; they contain most of the transcriptional control signals of the virus, including promoters and polyadenylation signals. Circular proviral DNA is found in the nuclei of infected cells and is thought to be the form which integrates into cellular DNA. Integration occurs at specific sites on the proviral DNA and thus the integrated and unintegrated forms are colinear

(Fig. 7c). Expression of the viral genome occurs by transcription of the integrated proviral DNA. The positioning within the LTRs of the viral promoter and polyadenylation signal ensures that the primary transcript is identical to virion RNA and this species can subsequently be packaged. Splicing generates subgenomic mRNAs which are translated to produce the products of the 5′-distal genes.

Strongly transforming retroviruses have a different genomic structure. Figure 7d shows the genome of Rous sarcoma virus (RSV), which contains an extra gene, denoted *src*, which is responsible for the ability of the virus to transform fibroblasts *in vitro* and to induce sarcomas *in vivo*. RSV is unique amongst strongly transforming retroviruses in that its replication genes are intact. Figure 7e shows the general structure of such a virus in which the transforming gene, designated *onc* for oncogene, is inserted into the prototype leukosis virus genome in such a way as to disrupt one or more of the replication genes. Such viruses are therefore defective and must be propagated with a helper virus. The *onc* gene is expressed under the control of the viral promoter located in the left-hand LTR. In many cases the *onc* gene is inserted into the *gag* gene and is expressed as a *gag—onc* fusion protein.

In all cases so far analysed the *onc* genes of transforming retroviruses have been shown to have cellular homologues. These c-*onc* genes are strongly conserved throughout evolution and have generally been found to be transcribed in a variety of normal cells. In the case of c-*src* the corresponding protein has also been identified in uninfected cells. It thus appears that retroviruses are natural transducing agents which have acquired their *onc* genes by some sort of recombination event with either host DNA or, more likely, with host mRNA via reverse transcription. One presumes that these viruses can also acquire other cellular genes but that the *onc* genes are the only ones presently recognized because the strongly transforming retroviruses thus generated have such an easily detectable phenotype. Analysis of the transforming viruses has indicated how foreign genes should be inserted into retroviral genomes in order to ensure their efficient expression and thus the rules for using retroviruses as vectors are reasonably clear. The availability of cloned proviral DNA means that genomes can be manipulated, and foreign genes can be inserted, using the same strategies so effectively applied to SV40.

2 The use of retroviral vectors

Thus far only one report has appeared describing the use of a retrovirus as a vector. Wei *et al.* (1981) have used HaMSV as a vector for the tk gene of HSV. The tk gene was inserted into the cloned proviral

DNA of a deletion mutant of HaMSV so that the recombinant virus retains the ability to express its *onc* gene. Transfection of the recombinant genome into tk⁻ NIH3T3 cells leads to focus formation, because of the expression of the *onc* gene, and also to clones capable of growing in HAT medium because they are expressing the HSV tk gene. HaMSV is defective for replication and thus the transfectants do not produce virus; they contain the recombinant genome integrated into their chromosomal DNA. However, by infecting transfectant clones with a replication competent virus, Moloney murine leukaemia virus, it was possible to rescue the recombinant virus and thus to propagate it as a helper-dependent retrovirus. Cells infected by this virus express the HSV thymidine kinase enzyme, detected immunologically, and direct analysis of the genomic RNA of the rescued recombinant virus shows that it contains linked HaMSV and HSV tk sequences. This model experiment clearly demonstrates the power of retroviral vectors and we can expect to see considerable use of such systems. The experiments of Wei *et al.* are of particular interest because they incorporate into a retroviral genome a genetic marker that can be selected biochemically; although the model recombinants constructed retain the ability to express the *onc* gene of HaMSV, it should be possible to construct non-transforming derivatives which retain the ability to express tk. Thus by inserting into such a genome another foreign gene, the tk⁺ retrovirus can be used to transduce it into cells in the absence of morphological transformation. As discussed above, the tk selection is non-ideal because tk⁻ derivatives of many interesting cell types are not available. However, the work of Wei *et al.* clearly shows how it will be possible to construct retroviral genomes which express the dominant selectable markers discussed in Section V.

Retroviruses offer a number of distinct advantages over the viral vector systems previously discussed. One factor which limits the utility of many viral vectors is host range: SV40 replicates efficiently only in monkey cells, human adenoviruses only in human cells. Retroviruses, however, often have extremely wide host ranges and thus one vector can be used to deliver the gene of interest to many different cell types. This wide host range can be extended even further by a process known as pseudotyping. If two retroviruses infect the same cell the genome of one virus can be packaged into the coat of the other. Host range in the next cycle of infection is determined by the coat, so it is possible to use pseudotypes to introduce a recombinant genome into a cell type refractory to infection by the original vector retrovirus. The mechanism by which retroviruses integrate into the host genome is also advantageous. As shown in Fig. 7 the integrated provirus is colinear with the unintegrated form because the LTRs contain signals directing the specificity of integration. One can therefore be confident that a gene inserted between two retroviral LTRs will be integrated intact, thus avoiding the problems of rearrangement which often occur when DNA virus genomes, or the HSV tk gene, are used to integrate exogenous DNA into cellular chromosomes. Furthermore, retrovirus infection does not lead to cell death; cells infected with, and transformed by, a retrovirus will continue to produce infectious virions for many cell generations. The establishment of such producer cell lines carrying a recombinant retroviral genome may thus provide long term cultures continuously expressing the foreign gene. This ability to produce infectious virions without killing the culture is not available for DNA virus vectors such as SV40 and adenovirus. One particular retrovirus, mouse mammary tumour virus (MMTV), has a unique advantage. Transcription of MMTV proviral DNA is regulated by glucocorticoids. If the sequences regulating the hormone response are within the MMTV LTRs, it should be possible to use this virus as a vector so that the expression of the inserted DNA is hormonally controlled. Finally, exogeneous retroviruses can be introduced into the germ line of animals (Jaenisch *et al.*, 1981) and thus, as is discussed in Section VI, these viruses provide a promising system for the genetic engineering of organisms.

IV   Episomal vectors based on viral genomes

A   Introduction

The use of replication competent, infectious virions as vectors allows the amplification of the cloned gene within the infected cell and thus the production of large amounts of the RNA and protein products of that gene. However, in such systems (with the exception of retroviruses) infection leads to the death of the host cell and thus a separate infection is required for each experiment. It would be highly desirable to develop systems in which the viral vector replicates as an episome, so that cell lines which continuously replicate and express the exogenous gene without the production of infectious virions could be established. Moreover, such systems would circumvent the limits imposed by the amount of DNA that can be packaged into a virus particle.

B   Papillomavirus vectors

The morphological transformation of cells by most DNA tumour viruses, and by retroviruses, is accompanied by the covalent integration of the viral genome into the chromosomal DNA of the host cell. However, recent work has shown that in cells transformed *in vitro* and in tumours induced *in vivo* by bovine papillomavirus (BPV) the viral genome is maintained as an episome and covalently integrated viral sequences are not detectable (Law *et al.*, 1981; Moar *et al.*, 1981a, b). Papillomaviruses are members of the same family, the papovaviruses, as SV40 and polyoma, but have been much less extensively characterized because there is no cell culture system in which infectious virions can be propagated. These viruses are widely distributed in nature, causing warts and a variety of other, generally benign, fibromas. Until recently virions and viral DNA could be obtained only by extraction of infected tissue. Now, however, the use of standard recombinant DNA technology allows the production of large amounts of viral DNA in *E. coli* and thus it will be possible to determine the structures of the viral genomes at the nucleotide sequence level. At present we do not know the details of the arrangement of the viral transcription units and so it is not possible to design vectors with the precision applicable to SV40 and adenoviruses.

Lowy *et al.* (1980) have shown that a defined, subgenomic fragment comprising 69% of the BPV-1 genome retains the ability to transform mouse cells *in vitro*. Sarver *et al.* (1981a, b) have used this fragment as a vector to introduce the rat preproinsulin gene into cultured mouse cells using morphological transformation as the selective marker. The transformed cells maintain the recombinant genome as an episome, each cell containing approximately 100 copies. The rat preproinsulin gene is efficiently expressed in such transformed cells; the preprotein is processed to proinsulin which is secreted into the medium but, as in the case of the SV40 rat preproinsulin recombinants (see Section III.A.4), the mature hormone is not produced. Endonuclease S1 mapping of the preproinsulin transcripts shows that they have the same 5′ termini as authentic preproinsulin mRNA from rat insulinomas, suggesting that transcription is initiated at the preproinsulin gene's promoter rather than from a BPV promoter.

Although this is the only published case of the use of BPV as a vector the system has great promise. The recombinant molecules replicate episomally, there is no integration and no infectious virions are produced. It should therefore be possible to maintain and express a gene of any size in permanent cell lines. As our knowledge of the structures of papillomavirus genomes increases, it will be possible to design vectors with more precision. Such information will also allow the incorporation into papillomavirus vectors of selectable genes (see Section V).

C   Episomal vectors based on SV40

1   *Semi-permissive cell systems*

SV40 DNA replicates episomally in transformed semi-permissive cells, e.g. human cells. However, such transformed cells always contain integrated viral genomes as well and the episomal forms are usually defective, containing extensive rearrangements of viral sequences. Upcroft *et al.* (1978) have described the transformation of rat and monkey cells by a recombinant SV40 genome carrying an *E. coli* tRNA gene. In the transformants the recombinant replicates as an episome, which is surprising in the case of the rat cells which are generally thought to be non-permissive. Because of the occurrence of integration and rearrangement this type of system has found little application. However, the SV40 genome has formed the basis of three other types of episomal vector.

2   *The replication and expression in COS cells of plasmids containing the SV40 origin*

COS cells contain functional SV40 large T-antigen and thus complement early mutants of the virus (see Section III.A.5). Because the only *cis*-acting function required for SV40 DNA replication is a segment of approximately 85 bp surrounding the SV40 origin of DNA replication, COS cells should replicate any DNA molecule containing this sequence. Myers and Tjian (1980) have constructed a series of plasmids which contain this small segment of the viral genome and shown that they replicate when transferred into COS cells. However, the efficiency of replication depends critically on the plasmid vector used. When the vector is pBR322, recombinants recovered from transfected monkey cells are frequently observed to have suffered deletions of plasmid sequences (Lusky and Botchan, 1981). These deleted derivatives replicate much more efficiently than the parental plasmids when they are transfected into monkey cells containing T-antigen. It appears that pBR322 contains sequences, located between the tetracycline resistance gene and the plasmid origin of DNA replication, which inhibit replication in animal cells. In transfection experiments of this type it is therefore important to use plasmid vectors from which this "poison sequence" has been removed. In addition to the plasmids described by Myers and Tjian (1980) and Lusky and Botchan (1981) the following are suitable: pSV08 and pSV010 (Learned *et al.*, 1981); pBR327 and pBR328 (Soberon *et al.*, 1980); pAT153 (Twigg and Sherratt, 1980) and pXf3 (D. Hanahan, personal communication). These plasmids are all deletion derivatives of pBR322 lacking the *nic/bom* site involved in plasmid mobilization. The poison sequence is located in the same region of the plasmid DNA as this mobilization site although it is not clear whether this coincidence is functionally significant.

One can therefore construct recombinants in which the gene of interest is cloned into a suitable plasmid vector containing the SV40 origin, transfect this DNA into COS cells and efficiently amplify the recombinant genome. Mellon *et al.* (1981) have constructed such recombinants containing the human α-globin gene. T-antigen-mediated DNA replication leads to the accumulation of 200 000—400 000 copies of the recombinant genome per transfected cell and the α-globin gene is efficiently expressed from such amplified templates. Endonuclease S1 mapping experiments show that the globin mRNA produced has the same 5′ and 3′ termini as authentic α-globin mRNA and that the primary transcript is correctly spliced. The general applicability of this system is not yet clear; e.g. the human β-globin gene is much less efficiently expressed (Mellon *et al.*, 1981).

At present it seems that this system does not lead to the establishment of permanent cell lines expressing the exogenous gene. The precise reasons for this are unclear, but it is likely that the cells can not tolerate the presence of such high levels of replicating, extra-chromosomal DNA. In this regard the much lower copy number of replicating BPV episomes is likely to be advantageous. A further limitation on the system is the efficiency with which the DNA can be introduced into the cells, but in all probability protoplast fusion techniques (Schaffner, 1980; Rassoulzadegan *et al.*, 1982) will obviate this problem. Because this system allows such great amplification of the template, and therefore of its products, it will be of great value in the analysis of the mechanisms of transcription and will be widely employed to assay mutations generated *in vitro* which alter potential transcriptional control regions and post-transcriptional processing signals.

3   The enhancing effect of the SV40 72 bp repeat

The control region surrounding the origin of replication of the SV40 genome has recently been shown to have a quite remarkable property. The early promoter of SV40 comprises not only the "TATA" box and the cap sites but also an upstream sequence which contains a tandem repeat of 72 bp. The integrity of one copy of this tandem duplication is absolutely required for the expression of viral early functions (Benoist and Chambon, 1981; Gruss *et al.*, 1981a). Banerji *et al.* (1981) have shown that this 72 bp repeat enhances the expression of the rabbit β-globin gene in transfection experiments. If the cloned gene alone is transfected into HeLa cells none of the cells expresses β-globin, detected immunologically, whereas if the recombinant also contains a segment of SV40 DNA including the 72 bp repeat approximately 1% of the cells express β-globin. This difference in expression could also be detected by endonuclease S1 mapping of the β-globin transcripts. The effect is not caused by an increased copy number due to replication from the SV40 origin; the enhancement works just as well when the plasmid contains the corresponding segment of SV40 DNA from a mutant with a defective origin. The effect is *cis*-acting, independent of the relative orientations of the gene and the 72 bp repeat and can be propagated over quite large distances, at least several kilobase pairs.

Several other reports have now appeared in which this enhancer effect has been utilized in studies on the expression of transfected genes. Grosveld *et al.* (1982) have studied the role of flanking DNA sequences in the transcription of the rabbit β-globin gene while Monschonas *et al.* (1981) and Busslinger *et al.* (1982) have used the system to analyse the expression of cloned human $β^0$- and $β^+$-thalassaemia genes.

The mechanism by which the SV40 72 bp repeat enhances the expression of adjacent genes is obscure but the technical utility of the effect is already proven. It would appear that such sequences are not restricted to the SV40 genome. Levinson *et al.* (1982) have shown that insertion into an SV40 genome lacking both copies of the repeat of a 72 bp tandem duplication from the LTR of Moloney murine sarcoma virus (Mo-MSV) reactivates the SV40 genome. The Mo-MSV repeat has no detectable sequence homology to the SV40 repeat but can assume its function. Moreover, Conrad and Botchan (1981) have observed that a segment of human DNA, originally isolated because it has sequence homology to the SV40 origin, has an enhancing effect on the efficiency of transformation by the HSV tk gene. Prior to the observation of the direct effect of the 72 bp repeat on globin gene expression, Capecchi (1980) showed that the frequency of tk transformation was increased if the recombinant plasmid contained sequences from the origin region of SV40 or, strikingly, from the LTR of RSV. Capechhi's interpretation of his data was that the viral sequences facilitate integration of the transforming DNA but with hindsight it seems likely that he was also observing the enhancing effect of 72 bp repeats. There now exist derivatives of many plasmid (Myers and Tjian 1980; Learned *et al.*, 1981; Lusky and Botchan, 1981) and cosmid (W. Chia, M. R. D. Scott and P. W. J. Rigby; F. G. Grosveld, T. Lund, A. L. Mellor, H. H. M. Dahl, H. Bullman, H. Budd and R. A. Flavell, both unpublished data.) vectors which carry the relevant region of the SV40 genome and one can expect to see such vectors routinely employed in primary cloning experiments so that subsequent analyses of gene expression are facilitated.

4   Plasmid vectors incorporating SV40 transcription units

Berg and his colleagues have developed a series of vectors which contain (i) an *E. coli* plasmid vector derived from pBR322; (ii) an SV40 transcription unit, generally the early unit, modified so that exogenous genes can be inserted in such a way that their transcription is guaranteed; (iii) in some cases an intact early region, derived from either SV40 or polyoma, to provide both large T-antigen and a replication origin so that the recombinant genome can replicate in permissive cells and (iv) a gene which can be selected for in all cell types. The construction and characterization of these vectors is described by Berg (1981) and by Mulligan *et al.* (1982). Thus far the major use of these vectors has been in the development of the dominant selectable systems discussed in Section V.

Figure 8 shows the structures of a representative series of these vectors. The first class is exemplified by pSV1GT5. The DNA of the late replacement vector SVGT5 has been cloned, via its unique *Pst*I site, into the SV40 *Pst*I site of a plasmid, pSV1, containing the segment of SV40 DNA from the *Bam*I site to the *Hind*III site at nucleotide 3476. This plasmid is routinely manipulated with an exogenous gene, X occupying the insertion site of SVGT5. The availability of this plasmid means that all of the manipulations involved in the construction of additional SVGT5 recombinants can be performed in *E. coli*, thus saving much time and expense. Once the desired construct has been produced and verified the late replacement vector recombinant SVGT5-X can be excised by *Pst*I digestion. Alternatively, because pSV1GT5 contains a duplication of the viral sequences between the *Hind*III site at 3476 and the *Bam*I site, transfection of pSV1GT5-X recombinant DNA into permissive monkey cells will lead to the excision of SVGT5-X by homologous recombination. In the analogous vector pSV1GT7 the inserted gene is transcribed into 19S mRNAs.

pSV2-X is the prototype of the second series of vectors. It contains the segment of pBR322 stretching anticlockwise from the *Eco*RI site to the *Pvu*II site, which includes the ampicillin-resistance gene and the plasmid origin of DNA replication. Adjacent to the *Pvu*II site is a segment of SV40 DNA containing the early promoter and the origin of viral DNA replication. This segment is bounded by the *Hind*III site at SV40 nucleotide 5171; thus this section of the recombinant does not contain any T-antigen coding sequences. The exogenous gene, X, is inserted between this *Hind*III site and a *Bgl*II site; this latter site is followed by two segments of SV40 DNA.

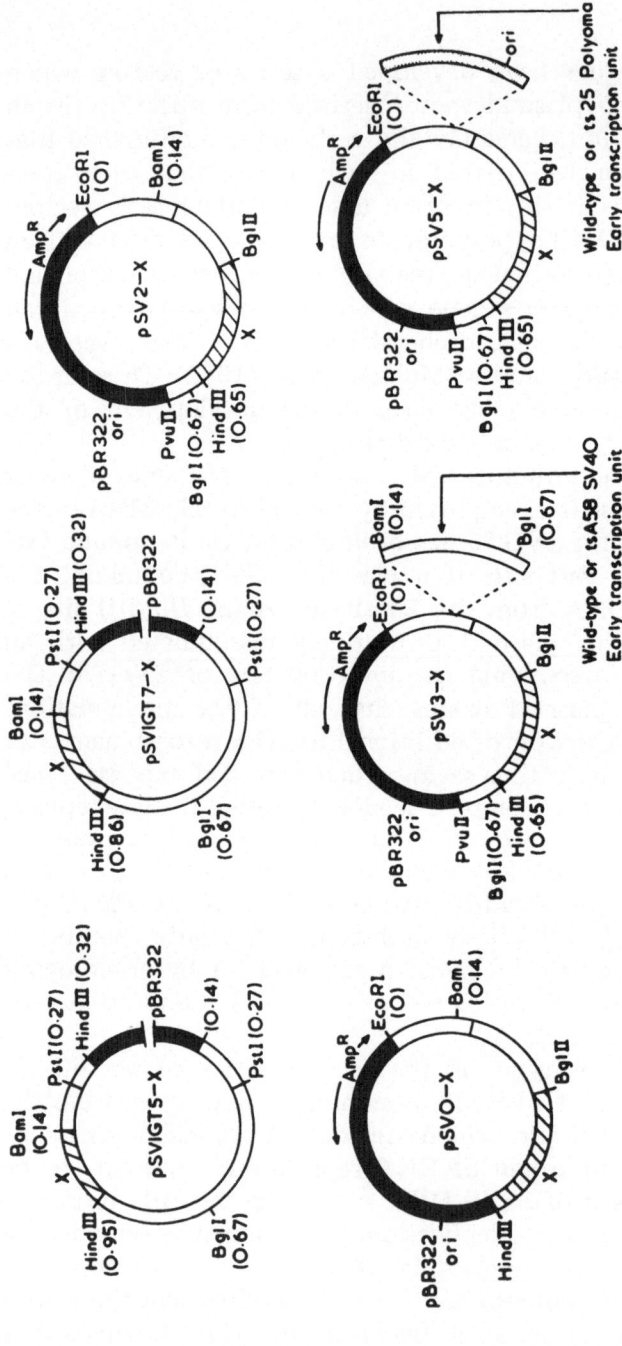

Figure 8 Plasmid vectors based on Simian virus 40. Open segments are derived from SV40; features of the viral genome are defined in map units. Solid segments are derived from pBR322, stippled segments from polyoma virus and the cross-hatched segments, X, indicate the location of the exogenous gene. pSV1GT5 and pSV1GT7 are cloned late replacement vectors from which complementable, infectious viral genomes can be derived. pSV3 and pSV5 are replicating vectors but are designed for episomal propagation or integration. pSV0 and pSV2 are non-replicating vectors. The structures are schematic and are not drawn to scale. Data taken from: Mulligan and Berg (1980, 1981a, b); Berg (1981); Subramani *et al.* (1981); Mulligan *et al.* (1982).

The first contains the small t-antigen intron, the second contains the viral early polyadenylation signal. Thus, transcription initiating at the SV40 early promoter will read through the foreign gene, will then traverse the small t-antigen intron, thus endowing the primary transcript with a functional splicing site, and finally will reach the polyadenylation site. The production of mature mRNA including the exogenous gene can thus be assured. Further derivatives of pSV2 contain one of the selectable genes discussed in Section V.B.

pSV0-X is derived from pSV2-X by excision of the *Pvu*II/*Hind*III fragment containing the SV40 promoter. Insertion of foreign DNA at the unique *Hind*III site of this vector can be used to assay for eukaryotic promoter sequences capable of initiating transcription of the gene X, which in such cases is likely to be one of the selectable genes discussed in Section V.

pSV3-X and pSV5-X are derived from pSV2-X by the insertion at the *Bam*I site of a complete SV40 (pSV3) or polyoma (pSV5) early region which includes the entire large T-antigen coding sequence and a functional viral origin of DNA replication. pSV3-X vectors should thus replicate in monkey cells and pSV5-X vectors in mouse cells. Mulligan *et al.* (1982) present evidence that such replication does occur, but it is relatively inefficient, presumably because the pBR322 moiety of these vectors retains the poison sequence. There also exist derivatives of pSV3 and pSV5 containing, respectively, the early region of SV40 mutant *tsA*58 and the early region of the polyoma *tsa* mutant ts25. These mutant T-antigens are incapable of initiating viral DNA replication at the non-permissive temperature and thus these mutant vectors can be manipulated to permit or abolish DNA replication. Moreover, the mutant T-antigens can not autoregulate their own synthesis at the non-permissive temperature and thus one might expect to be able to achieve higher levels of transcription from the viral early promoter.

V Virus-based vectors carrying selectable genes

A Introduction

A final approach to the use of the SV40 genome as a vector is exemplified by the recent work of Berg and his colleagues. They have concentrated upon the development of plasmid vectors which are propagatable in *E. coli*, replicate in animal cells using the SV40 (or polyoma) origin and large T-antigen, do not pass through the viral life cycle and are therefore not subject to packaging constraints and which are selectable in all cell types. The ability to select for cells carrying the recombinant genome is clearly of paramount importance. DNA transfection techniques lead to the uptake of the DNA by only a relatively small percentage of the cells and to screen large numbers of transfectants for the presence of the transfecting DNA by, for example, hybridization techniques, is unacceptably tedious. Although it now appears that protoplast fusion may allow the DNA to be introduced into all of the cells in a population it is not clear with what efficiency this method leads to stable expression. Moreover, the fusion must still be repeated for each experiment and there may be many experiments, in which one is looking for a biological effect on the cells, in which the presence of bacterial debris will be unacceptable. The desirability of selectable systems thus remains.

The tk gene of HSV has been widely used as a selectable marker but suffers from the great disadvantage that it can only be used in conjunction with tk⁻ mutant cells. Such mutants are not available for many cell types, and the derivation of such mutants is not a trivial proposition. tk⁻ mutants that can be efficiently transfected are rare, the only two well documented systems being mouse Ltk⁻ cells (Wigler *et al.*, 1979) and the Rat-2 line of fibroblasts developed by Topp (1981). Berg's group have therefore sought to develop dominant selectable markers that can be used in any cell type.

B   The *gpt* system

The first selection system that they have developed employs the *E. coli gpt* gene which encodes the enzyme xanthine—guanine phosphoribosyltransferase (XGPRT), which is the bacterial analogue of the mammalian enzyme hypoxanthine—guanine phosphoribosyltransferase (HGPRT). The bacterial enzyme will efficiently convert xanthine into XMP, which is a precursor to GMP, whereas the mammalian enzyme will utilize only hypoxanthine and guanine. Mulligan and Berg (1980) cloned the *E. coli gpt* gene into the SV40 late replacement vector SVGT5 (see Section III.A.4) and propagated the recombinant as infectious virions with a *tsA* mutant helper. CV-1 cells infected with the mixed virus stock synthesize considerable amounts of *E. coli* XGPRT, which can be distinguished from the endogenous mammalian HGPRT by virtue of its insensitivity to inhibition by hypoxanthine and its electrophoretic mobility. They also constructed derivatives of pSV2, pSV3 and pSV5 in which the exogenous gene X is *gpt*. Irrespective of whether replicating (pSV3) or non-replicating (pSV5) vectors were used, monkey cells transfected with the *gpt* recombinants synthesized the *E. coli* XGPRT enzyme. The Lesch—Nyhan syndrome results from a defect in

HGPRT; human skin fibroblasts derived from patients with the syndrome and immortalized by SV40 transformation are therefore incapable of growth in HAT medium. Transfection of such cells with pSV-*gpt* DNA leads to biochemical transformation and the growth of clones which can multiply in HAT medium. Thus the expression of the *E. coli gpt* gene under the influence of SV40 transcriptional control signals is capable of overcoming the Lesch—Nyhan genetic defect. As expected, the biochemically transformed cells express the *E. coli* enzyme and it appears that this expression is genetically stable in the absence of selective pressure.

The ability to express the *E. coli* XGPRT in mammalian cells allowed Mulligan and Berg (1981a) to develop a positive selection system. Figure 9 shows the steps in purine metabolism relevant to this selection. Mycophenolic acid is an inhibitor of IMP dehydrogenase which blocks the conversion of IMP into XMP and thus the *de novo* synthesis of GMP. This inhibition can be made more effective by also adding aminopterin to the medium, which blocks the synthesis of IMP. Mammalian cells can grow in the presence of these two inhibitors if the medium is further supplemented with guanine and either hypoxanthine or adenine; the salvage pathway phosphoribosyltransferases can convert these bases to the corresponding mononucleotides. The inhibition is not overcome if the medium is supplemented with adenine and xanthine because the mammalian HGPRT can not convert xanthine to XMP and thus GMP can not be produced. If, however, the mammalian cells were expressing the *E. coli* XGPRT, growth on adenine plus xanthine would be possible.

Mulligan and Berg transfected both monkey and mouse cells with pSV series plasmid vectors containing the *gpt* gene and showed that

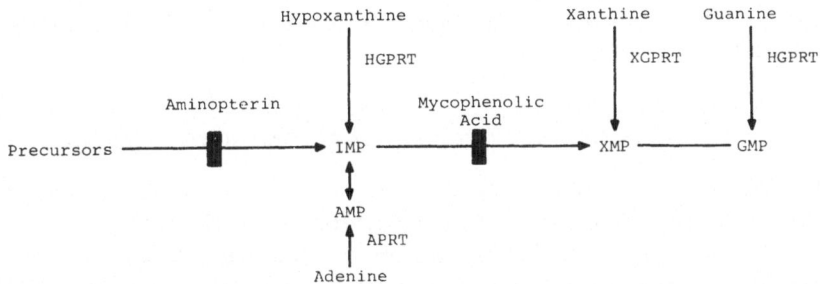

Figure 9   Purine metabolism relevant to the *gpt* selection system. HGPRT is the mammalian hypoxanthine—guanine—phosphoribosyltransferase. XGPRT is the *E. coli* xanthine—guanine—phosphoribosyltransferase. APRT is the mammalian adenine—phosphoribosyltransferase. For further details see Mulligan and Berg (1981a).

when grown in the selective medium the transfected cells gave rise to 2—25 clones per 10^5 cells. Transfection with the same vectors containing rabbit β-globin cDNA instead of *gpt* gave no growth. The transfectants synthesize *E. coli* XGPRT as judged by its electrophoretic mobility and insensitivity to inhibition by hypoxanthine. Some of the recombinants used in these studies (e.g. pSV3-*gpt*) contain a complete SV40 early transcription unit. It is therefore possible to ask whether clones selected for the expression of *gpt* are also expressing the non-selected marker. Several clones, derived from both monkey and mouse cells, were shown to be synthesizing small t-antigen, but authentic large T-antigen was not detected. It thus appears likely that the *gpt* selection can be used to construct cells expressing a non-selected gene and Lee *et al.* (1981), in a series of experiments to be discussed below, have provided further evidence that this is the case. The biochemically transformed clones selected in this way contain the transforming DNA integrated into chromosomal DNA; episomal replication was not detected. This selection system fulfills the criteria discussed above; it is genetically dominant and can therefore be applied to any cell type.

Mulligan and Berg (1981b) have further investigated the expression of the *gpt* gene in transduced mammalian cells. Endonuclease S1 mapping experiments show that in general the *gpt* gene is transcribed from the SV40 promoters and that the primary transcripts are processed as one would expect. However, in cells transfected with pSV1GT5-*gpt* and pSV1GT7-*gpt* aberrant splicing, in which the acceptor site is located within the prokaryotic DNA segment, was detected. Similar aberrant splicing has been observed in SVGT7-β-globin cDNA recombinants; in this case the acceptor is located within the 5'-untranslated segment of the cDNA (Mulligan *et al.* 1980). It is thus clear that even with the most careful vector construction it is necessary to check the structures of the mRNAs produced in transduced cells.

C   The aminoglycoside phosphotransferase system

An alternative dominant selection system employs the gene for a bacterial aminoglycoside 3' phosphotransferase; this gene originates from the transposon Tn5 and confers on bacterial strains expressing it resistance to neomycin and kanamycin. Jiminez and Davies (1980) showed that yeasts are sensitive to G-418, a 2-deoxystreptamine antibiotic which inhibits eukaryotic protein synthesis. If, however, yeast cells are transfected with Tn601 before exposure to the antibiotic, G418-resistant clones arise which can be shown to be expressing the phosphotransferase encoded by Tn601. Colbere-Garapin *et al.* (1981) have extended this work by showing that the Tn5 phosphotransferase gene can form the basis of a dominant selection system for higher eukaryotic cells. They cloned the phosphotransferase gene into a plasmid containing the tk gene of HSV in such a way that the bacterial gene could be expressed from the tk gene's promoter. Further manipulations refined the construct so that the bacterial gene was positioned closer to both the tk promoter and the tk polyadenylation site. Transfection of mammalian cells with such a plasmid leads to the occurrence of G418-resistant clones which can be shown to be expressing the bacterial enzyme. The transfected sequences appear to be integrated into chromosomal DNA. If the phosphotransferase plasmid is co-transfected into Ltk⁻ cells together with a plasmid containing the HSV tk gene, 45% of the G418-resistant clones were found to be tk⁺, indicating that selection for G418-resistance can be used to isolate clones carrying and expressing a non-selected marker.

This same Tn5 phosphotransferase gene has also been exploited by Berg's group. They have constructed a series of pSV vectors, analogous to the *gpt* vectors discussed above, which express the phosphotransferase gene under SV40 control and shown that such vectors can also be used positively to select transfected cells (Berg, 1981).

D   The dihydrofolate reductase system

A further selection system employs the *dhfr* gene, although this selection is not dominant, being analogous to the tk system. Subramani *et al.* (1981) have cloned mouse *dhfr* cDNA in both SV40 late replacement vectors and in pSV series plasmid vectors. Infection of permissive monkey cells with SVGT5- and SVGT7-based *dhfr* recombinants leads to the synthesis of large amounts of the mouse protein in the same way previously described for analogous β-globin and *gpt* recombinants. DHFR-negative CHO cells can not synthesize tetrahydrofolate and therefore require growth medium supplemented with thymidine, glycine and purines. Transfection of these mutant cells with pSV2-*dhfr* gives rise to clones able to grow in the absence of supplements. The transfected cells have a level of DHFR activity twice that of wild-type cells.

Although this selection system is not dominant it does have one notable advantage. Wild-type cells are killed by the antifolate drug methotrexate. Cells resistant to methotrexate can be selected by growth in the presence of increasing concentrations of the drug and Schimke *et al.* (1981) have shown that the predominant cause of resistance is amplification of the *dhfr* gene. However, the amplification unit is large, involving extensive regions of DNA flanking the *dhfr* gene itself. Thus, by transfecting DHFR-negative cells with a recombinant containing the gene of interest linked to the *dhfr* gene and then selecting for methotrexate resistance it should be possible to amplify the linked gene. Kaufman *et al.* (1982) have successfully employed just this strategy. They constructed a plasmid in which *dhfr* cDNA is joined to the splicing signals of an immunoglobulin gene and also to an intact SV40 genome. They then transfected this recombinant into DHFR-negative CHO cells, selected for DHFR expression and then selected methotrexate resistance and thus for amplification. Some of the methotrexate-resistant clones that they isolated were found to be expressing SV40 gene products at high levels, presumably because the viral genome had also been amplified. In one clone, SV40 small t-antigen accounted for 2% of the total soluble protein. This system remains relatively uncharacterized but it may well provide a powerful means of constructing cell lines which express non-selectable genes large amount from chromosomally integrated sequences.

O'Hare *et al.* (1981) have described another variation on the use of *dhfr* as a selectable gene. They constructed plasmids containing the gene for a bacterial DHFR intrinsically resistant to methotrexate. The bacterial gene is positioned so that it should be transcribed from the SV40 early promoter and the plasmid also carried a segment of the rabbit β-globin gene containing a splice junction and a poly-adenylation signal. Transfection of this plasmid into mouse cells led to the occurrence of methotrexate-resistant clones which contain integrated plasmid DNA and synthesize mRNA of the expected size with 5′ ends which map to the SV40 early promoter. This system should allow the construction of cell lines expressing non-selected genes but this point has not yet been tested.

Lee *et al.* (1981) have reported a particularly interesting application of the pSV2-*dhfr* plasmid described by Subramani *et al.* (1981). In pSV2-*dhfr* the mouse *dhfr* cDNA is expressed under the control of the SV40 early promoter and this plasmid will transform DHFR-negative CHO cells, allowing growth in medium lacking thymidine, glycine and purines. Lee *et al.* replaced the segment of DNA carrying the SV40 promoter with a segment of the LTR of MMTV which is thought to carry the hormone-responsive viral promoter. Transfection with this pMTV*dhfr* plasmid also leads to biochemical transformation although the frequency is considerably lower than that obtained with pSV2*dhfr*. This lower frequency may merely reflect the fact that the MMTV promoter is weaker than the SV40 promoter. They therefore constructed another plasmid, pMDSG, by inserting into pMTV*dhfr* a segment of DNA containing the *gpt* gene together with the SV40 sequences necessary for the efficient expression of this bacterial gene in eukaryotic cells. This plasmid allows transformants to be selected for *gpt* expression; clones selected in this way grow as well in the DHFR-selective medium as clones selected directly for *dhfr* expression. However, if pMDSG-transfected cells are selected directly for *dhfr* expression the frequency is as high as with pSV2*dhfr*. This is probably due to the enhancing effect of the SV40 72 bp repeats carried on the *gpt* segment (see Section IV.C.3). By measuring the sensitivity of the transformed cells to methotrexate, and by measuring the amount of DHFR protein produced, Lee *et al.* could show that expression of the mouse *dhfr* cDNA from the MMTV promoter is regulated by the synthetic glucocorticoid dexamethasone. In agreement with this, endonuclease S1 mapping experiments showed that the 5′ ends of the *dhfr* mRNA map to the segment of the MMTV LTR thought to contain the viral promoter.

As discussed above (Section III.C.2) transcription of MMTV proviral DNA is known to be regulated by dexamethasone. The importance of the work of Lee *et al.* is that it shows clearly that the transcription of exogenous genes fused to the MMTV LTR can also be hormonally regulated. This effect is likely to be general as Lee *et al.* discuss data indicating that expression of *E. coli* XGPRT is regulated by dexamethasone when the *gpt* gene is joined to the MMTV LTR. The MMTV promoter appears to be intrinsically rather weak but transcription from it can be increased by a defined hormonal stimulus. This system may therefore be of great value for studying genes which are lethal to the cell if expressed at high levels. There are many examples of *E. coli* genes, e.g. that for DNA polymerase I, which can not be cloned in high copy number plasmids because high level expression is lethal and it would be surprising if there are not analogous eukaryotic genes. It should be possible to construct cell lines carrying such genes fused to the MMTV LTR so that the gene is expressed at sub-lethal levels; the expression of such genes can then be analysed in subcultures hormonally induced to high levels of expression. Constructs in which the MMTV LTR is fused to genes with easily assayable products should prove invaluable in experiments to dissect the interaction of the glucocorticoid receptor with the viral promoter. Such systems will be analogous to the *lac* fusions which have been invaluable in the analysis of *E. coli* promoters.

VI Vectors for the integration of exogenous DNA into chromosomal DNA

The fact that most tumour viruses integrate their genomes into host chromosomal DNA during transformation can be exploited to introduce exogenous DNA into the chromosomes of cultured cells. It was originally thought that tumour viruses would provide the vectors of choice for the covalent integration of exogenous genes, but we now know that many other kinds of DNA will integrate; both tk co-transformation and the series of selectable vectors developed by Berg's group lead to covalent integration of the exogenous gene. However, morphological transformation by tumour viruses does provide another selectable character. Transformation by SV40 requires the expression of only the early region of the viral genome (Tooze, 1980) and thus if the late genes are replaced by a segment of foreign DNA and the recombinant is used to transform cells the foreign DNA should become integrated. Muzycka (1979) has shown that this is so when the foreign DNA is a segment of λ and Schaffner *et al.* (1979) have performed similar experiments with sea urchin histone genes. This approach is, however, fraught with difficulties. The recombination between viral and cellular DNAs usually occurs within the late region of the viral genome and thus it is likely that the foreign DNA segment will be disrupted. Moreover, as is discussed in Section III.A.2., the integration of SV40 DNA is accompanied by complex and continuing amplification and rearrangement events. It is therefore likely to be extremely difficult to predict, and impossible to control, the way in which the cloned sequences are finally arranged and expressed. It is a measure of the extraordinarily rapid progress achieved in this field in the past two years that what was recently thought to be the most obvious approach to the introduction of exogenous genes into mammalian cells should now be almost totally outmoded.

The fact that retroviruses integrate specifically via their LTRs (see Section III.C.I) means that they provide the most promising system for controlled integration but thus far this possibility has not been exploited. Moreover, the work of Jaenisch and his colleagues (Jahner and Jaenisch, 1980; Jaenisch *et al.*, 1981) has shown that it is possible by injecting retroviruses into mouse embryos which are then reimplanted into a pseudo-pregnant foster mother, to introduce an exogenous retrovirus into the germ line of the mouse. Because the expression of some retroviruses is tissue-specific, even when all cells of the animal carry integrated proviral DNA, it may be possible to exploit this system to achieve tissue-specific expression of the exogenous gene. Although recent work (Wagner *et al.*, 1981) has indicated that it may be possible to introduce cloned genes into the germ line of the mouse without recourse to viral vectors, the use of retroviral vectors for this purpose is likely to remain important.

VII   Conclusions

The last few years have seen the rapid development of a wide variety of systems for the reintroduction of cloned genes into cultured eukaryotic cells. Each of these systems has its advantages and disadvantages; e.g. vectors that allow high level expression of the exogenous gene are unlikely to be suitable for the establishment of permanent cell lines expressing that gene. Therefore a series of experiments designed to study one gene may well require the use of several different vectors. The technical development of eukaryotic vectors remains very rapid and the knowledge that we acquire by using these systems itself accelerates the rate of development. It is reasonable to expect that we shall shortly have available a set of vectors which allow any gene to be introduced efficiently into all cell types so that the level of expression of the foreign gene can be conveniently manipulated. Moreover, the first steps in extending this technology to the genetic engineering of organisms have already been taken.

The eukaryotic cloning vectors I have described are a powerful addition to the technical armoury of the eukaryotic molecular biologist. We now have the capability to isolate genes, manipulate and mutate them and then return them to their natural environment so that we can study their expression and their biological effects. Future developments will surely include the construction of mini-chromosomes which segregate properly at cell division and, as our knowledge of chromatin structure increases, the incorporation into such vectors of sequences responsible for conferring both nucleosomal and higher order structures. It is now possible to contemplate the construction of a small, artificial chromosome having a predetermined set of properties. The impact of such genetic constructions upon our understanding of the molecular bases of growth and differentiation will surely be immense.

VIII   Acknowledgements

I am grateful to Saveria Campo, Dick Flavell, Mary-Jane Gething, Peter Howley, George Khoury, Tom Maniatis, Richard Mulligan, Walter Schaffner, Phil Sharp, Bob Tjian, Charles Weissmann and Shermann Weissman for sending me preprints of their unpublished papers and to Sue Hayman for her excellent preparation of the manuscript. I am supported by a Career Development Award from the Cancer Research Campaign.

134   *P. W. J. Rigby*

IX References

Alwine, J. C., Kemp, D. J. and Stark, G. R. (1977). Method for detection of specific RNAs in agarose gels by transfer to diazobenzyloxymethyl-paper and hybridization with DNA probes. *Proc. Natn. Acad. Sci. U.S.A.* **74**, 5350—5354.

Banerji, J., Rusconi, S. and Schaffner, W. (1981). Expression of a β-globin gene is enhanced by remote SV40 DNA sequences. *Cell* **27**, 299—308.

Beggs, J. D., van den Berg, J., van Ooyen, A. and Weissmann, C. (1980). Abnormal expression of chromosomal rabbit β-globin gene in *Saccharomyces cerevisiae. Nature, Lond.* **283**, 835—840.

Bender, M. A. and Brockman, W. W. (1981). Rearrangement of integrated viral DNA sequences in mouse cells transformed by simian virus 40. *J. virol.* **38**, 872—879.

Benoist, C. and Chambon, P. (1981). *In vivo* sequence requirements of the SV40 early promoter region. *Nature, Lond.* **290**, 304—310.

Berg, P. (1981). Dissections and reconstructions of genes and chromosomes. *Science, N.Y.* **213**, 296—303.

Berk, A. J. and Sharp, P. A. (1977). Sizing and mapping of early adenovirus mRNAs by gel electrophoresis of S1 endonuclease-digested hybrids. *Cell* **12**, 721—732.

Botchan, M., Stringer, J., Mitchison, T. and Sambrook, J. (1980). Integration and excision of SV40 DNA from the chromosome of a transformed cell. *Cell* **20**, 143—152.

Breathnach, R., Mantei, N. and Chambon, P. (1980). Correct splicing of a chicken ovalbumin gene transcript in mouse L cells. *Proc. Natn. Acad. Sci. U.S.A.* **77**, 740—744.

Busslinger, M., Moschonas, N. and Flavell, R. A. (1982). β^+ thalassemia:aberrant splicing results from a single point mutation in an intron. *Cell* **27**, 289—298.

Capecchi, M. R. (1980). High efficiency transformation by direct microinjection of DNA into cultured mammalian cells. *Cell* **22**, 479—488.

Carbon, J., Shenk, T. E. and Berg, P. (1975). Biochemical procedure for production of small deletions in Simian virus 40 DNA. *Proc. Natn. Acad. Sci. U.S.A.* **72**, 1392—1396.

Chu, G. and Sharp, P. A. (1981). SV40 DNA transfection of cells in suspension: analysis of the efficiency of transcription and translation of T-antigen. *Gene* **13**, 197—202.

Clayton, C. E. and Rigby, P. W. J. (1981). Cloning and characterization of the integrated viral DNA from three lines of SV40-transformed mouse cells. *Cell* **25**, 547—559.

Colbere-Garapin, F., Horodniceanu, F., Kourilsky, P. and Garapin, A-C. (1981). A new dominant hybrid selective marker for higher eukaryotic cells. *J. Molec. Biol.* **150**, 1—14.

Conrad, S. E. and Botchan, M. R. (1981). Isolation and characterization of human DNA sequences homologous to the SV40 origin of replication. *J. Supramolec. Struct. Cell Biochem.* Suppl. **5**, 445.

Corden, J., Wasylyk, B., Buchwalder, A., Sassone-Corsi, P., Kedinger, C. and Chambon, P. (1980). Promoter sequences of eukaryotic protein-coding genes. *Science, N.Y.* **209**, 1406—1414.

DeLap, T. J., Rush, M. G., Zouzias, D. and Khan, S. (1978). Isolation and preliminary characterisation of the small circular DNA present in African green monkey kidney (BSC-1) cells. *Plasmid* **1**, 508—521.

Elder, J. T., Spritz, R. A. and Weissman, S. M. (1981). Simian virus 40 as a eukaryotic cloning vehicle. *A. Rev. Genet.* 15, 295—340.

Favaloro, J., Treisman, R. and Kamen, R. (1980). Transcription maps of polyoma virus-specific RNA: analysis by two-dimensional nuclease S1-gel mapping. *In* "Methods in Enzymology" (Eds L. Grossman and K. Moldave) Vol. 65, 718—749. Academic Press, New York.

Fitzgerald, M. and Shenk, T. (1981). The sequence 5'-AAUAAA-3' forms part of the recognition site for polyadenylation of late SV40 mRNAs. *Cell* 24, 251—260.

Ganem, D., Nussbaum, A. L., Davoli, D. and Fareed, G. C. (1976). Propagation of a segment of bacteriophage λ-DNA in monkey cells after covalent linkage to a defective simian virus 40 genome. *Cell* 7, 349—359.

Gething, M-J. and Sambrook, J. (1981). Cell-surface expression of influenza haemagglutinin from a cloned DNA copy of the RNA gene. *Nature, Lond.* 293, 620—625.

Gillam, S., Astell, C. R. and Smith, M. (1980). Site-specific mutagenesis using synthetic oligodeoxyribonucleotides: isolation of a phenotypically silent ∅X174 mutant, with a specific nucleotide deletion, at very high efficiency. *Gene* 12, 129—137.

Gluzman, Y. (1981). SV40-transformed simian cells support the replication of early SV40 mutants. *Cell* 23, 175—182.

Goff, S. P. (1977). Ph.D Thesis, Stanford University Medical School.

Goff, S. P. and Berg, P. (1976). Construction of hybrid viruses containing SV40 and λ phage DNA segments and their propagation in cultured monkey cells. *Cell* 9, 695—705.

Goff, S. P. and Berg, P. (1979). Construction, propagation and expression of Simian virus 40 recombinant genomes containing the *Escherichia coli* gene for thymidine kinase and a *Saccharomyces cerevisiae* gene for tyrosine transfer RNA. *J. Molec. Biol.* 133, 359—383.

Goldberg, M. (1978). PhD Thesis, Stanford University Medical School.

Goldenberg, C. J. and Raskas, H. J. (1981). *In vitro* splicing of purified precursor RNAs specified by early region 2 of the adenovirus 2 genome. *Proc. Natn. Acad. Sci. U.S.A.* 78, 5430—5434.

Graham, F. L. and van der Eb, A. J. (1973). A new technique for the assay of infectivity of human adenovirus 5 DNA. *Virology* 52, 456—467.

Graham, F. L., Smiley, J., Russell, W. C. and Nairn, R. (1977). Characteristics of a human cell line transformed by DNA from human adenovirus type 5. *J. Gen. Virol.* 36, 59—72.

Grodzicker, T. and Klessig, D. F. (1980). Expression of unselected adenovirus genes in human cells co-transformed with the HSV-1 tk gene and adenovirus 2 DNA. *Cell* 21, 453—463.

Grosveld, G. C., de Boer, E., Shewmaker, C. K. and Flavell, R. A. (1982). DNA sequences necessary for transcription of the rabbit β-globin gene *in vivo*. *Nature, Lond.* 295, 120—126.

Grummt, I. (1981). Specific transcription of mouse ribosomal DNA in a cell-free system that mimics control *in vivo*. *Proc. Natn. Acad. Sci. U.S.A.* 78, 727—731.

Gruss, P. and Khoury, G. (1980). Rescue of a splicing defective mutant by insertion of an heterologous intron. *Nature, Lond.* 286, 634—637.

Gruss, P. and Khoury, G. (1981). Expression of simian virus 40-rat preproinsulin recombinants in monkey kidney cells: Use of preproinsulin RNA processing signals. *Proc. Natn. Acad. Sci. U.S.A.* 78, 133—137.

Gruss, P., Lai, C-J., Dhar, R. and Khoury, G. (1979). Splicing as a requirement for biogenesis of functional 16S mRNA of simian virus 40. *Proc. Natn. Acad. Sci. U.S.A.* **76**, 4317—4321.

Gruss, P., Dhar, R. and Khoury, G. (1981a). Simian virus 40 tandem repeated sequences as an element of the early promoter. *Proc. Natn. Acad. Sci. U.S.A.* **78**, 943—947.

Gruss, P., Efstratiadis, A., Karathanasis, S., Konig, M. and Khoury, G. (1981b). Synthesis of stable unspliced mRNA from an intronless simian virus 40-rat preproinsulin gene recombinant. *Proc. Natn. Acad. Sci. U.S.A.* **78**, 6091—6095.

Gruss, P., Ellis, R. W., Shih, T. Y., Konig, M., Scolnick, E. M. and Khoury, G. (1981c). SV40 recombinant molecules express the gene encoding p21 transforming protein of Harvey murine sarcoma virus. *Nature, Lond.* **293**, 486—488.

Hamer, D. H. and Leder, P. (1979a). Expression of the chromosomal mouse β^{maj}-globin gene cloned in SV40. *Nature, Lond.* **281**, 35—40.

Hamer, D. H. and Leder, P. (1979b). SV40 recombinants carrying a functional RNA splice junction and polyadenylation site from the chromosomal mouse β^{maj}-globin gene. *Cell* **17**, 737—747.

Hamer, D. H. and Leder, P. (1979c). Splicing and the formation of stable RNA. *Cell* **18**, 1299—1302.

Hamer, D. H., Davoli, D., Thomas, C. A. and Fareed, G. C. (1977). Simian virus 40 carrying an *Escherichia coli* suppressor gene. *J. Molec. Biol.* **112**, 155—182.

Hamer, D. H., Smith, K. D., Boyer, S. H. and Leder, P. (1979). SV40 recombinants carrying rabbit β-globin gene coding sequences. *Cell* **17**, 725—735.

Hamer, D. H., Kaehler, M. and Leder, P. (1980). A mouse globin gene promoter is functional in SV40. *Cell* **21**, 697—708.

Henikoff, S., Tatchell, K., Hall, B. D. and Nasmyth, K. (1981). Isolation of a gene from *Drosophila* by complementation in yeast. *Nature, Lond.* **289**, 33—37.

Hiscott, J., Murphy, D. and Defendi, V. (1980). Amplification and rearrangement of integrated SV40 DNA sequences accompany the selection of anchorage-independent transformed mouse cells. *Cell* **22**, 535—543.

Hiscott, J. B., Murphy, D. and Defendi, V. (1981). Instability of integrated viral DNA in mouse cells transformed by simian virus 40. *Proc. Natn. Acad. Sci. U.S.A.* **78**, 1736—1740.

Hitzeman, R. A., Hagie, F. E., Levine, H. L., Goeddel, D. V., Ammerer, G. and Hall, B. D. (1981). Expression of a human gene for interferon in yeast. *Nature, Lond.* **293**, 717—722.

Hutchison, C. A., Phillips, S., Edgell, M. H., Gillam, S., Jahnke, P. and Smith, M. J. (1978). Mutagenesis at a specific position in a DNA sequence. *J. Biol. Chem.* **253**, 6551—6560.

Jaenisch, R., Jahner, D., Nobis, P., Simon, I., Lohler, J., Harbers, K. and Grotkopp, D. (1981). Chromsomal position and activation of retroviral genomes inserted into the germ line of mice. *Cell* **24**, 519—529.

Jahner, D. and Jaenisch, R. (1980). Integration of Moloney leukaemia virus into the germ line of mice: correlation between site of integration and virus activation. *Nature, Lond.* **287**, 456—458.

Jimenez, A. and Davies, J. (1980). Expression of a transposable antibiotic resistance element in *Saccharomyces*. *Nature, Lond.* **287**, 869—871.

Jones, N. and Shenk, T. (1978). Isolation of deletion and substitution mutants of adenovirus type 5. *Cell* **13**, 181—188.

Jones, N. and Shenk, T. (1979). Isolation of adenovirus type 5 host range deletion mutants defective for transformation of rat embryo cells. *Cell* 17, 683–689.

Kaufman, R. J., Latt, S. A. and Sharp, P. A. (1982). Expression and amplification of DNA introduced into mammalian cells. *In* "Gene Amplification" (Ed. R. T. Schimke). Cold Spring Harbor Laboratory, New York. In press.

Kozak, M. (1980). Evaluation of the 'scanning model' for initiation of protein synthesis in eukaryotes. *Cell* 22, 7–8.

Lai, C-J. and Khoury, G. (1979). Deletion mutants of simian virus 40 defective in biosynthesis of late viral mRNA. *Proc. Natn. Acad. Sci. U.S.A.* 76, 71–75.

Lai, C-J. and Nathans, D. (1974). Deletion mutants of Simian virus 40 generated by enzymatic excision of DNA segments from the viral genome. *J. Molec. Biol.* 89, 179–193.

Law, M-F., Lowy, D. R., Dvoretzky, I. and Howley, P. M. (1981). Mouse cells transformed by bovine papillomavirus contain only extrachromosomal viral DNA sequences. *Proc. Natn. Acad. Sci. U.S.A.* 78, 2727–2731.

Learned, R. M., Myers, R. M. and Tjian, R. (1981). Replication in monkey cells of plasmid DNA containing the minimal SV40 origin. *In* "Developmental Biology Using Purified Genes" (Eds D. Brown and C. R. Fox). ICN-UCLA Symposia on Molecular and Cellular Biology, Vol. 23. Academic Press, New York. In press.

Lebowitz, P. and Weissman, S. M. (1979). Organisation and transcription of the Simian virus 40 genome. *Curr. Topics Microbiol. Immunol.* 87, 43–172.

Lee, F., Mulligan, R., Berg, P. and Ringold, G. (1981). Glucocorticoids regulate expression of dihydrofolate reductase cDNA in mouse mammary tumour virus chimaeric plasmids. *Nature, Lond.* 294, 228–232.

Levinson, B., Khoury, G., Vande Woude, G. and Gruss, P. (1982). Activation of the SV40 genome by the 72 base-pair tandem repeats of Moloney sarcoma virus. *Nature, Lond.* in press.

Lowy, D. R., Dvoretzky, I., Shober, R., Law, M-F., Engel, L. and Howley, P. M. (1980). *In vitro* tumorigenic transformation by a defined sub-genomic fragment of bovine papilloma virus. *Nature, Lond.* 287, 72–74.

Luse, D. S. and Roeder, R. G. (1980). Accurate transcription initiation on a purified mouse β-globin DNA fragment in a cell-free system. *Cell* 20, 691–699.

Lusky, M. and Botchan, M. (1981). Inhibitory effect of specific pBR322 DNA sequences upon SV40 replication in Simian cells. *Nature, Lond.* 293, 79–81.

McCormick, F., Chaudry, F., Harvey, R., Smith, R., Rigby, P. W. J., Paucha, E. and Smith, A. E. (1980). T antigens of SV40-transformed cells. *Cold Spring Harb. Symp. Quant. Biol.* 44, 171–178.

Manley, J. L., Fire, A., Cano, A., Sharp, P. A. and Gefter, M. L. (1979). DNA dependent transcription of adenovirus genes in a soluble whole-cell extract. *Proc. Natn. Acad. Sci. U. S. A.* 77, 3855–3859.

Mantei, N., Boll, W. and Weissmann, C. (1979). Rabbit β-globin mRNA production in mouse L cells transformed with cloned rabbit β-globin chromosomal DNA. *Nature, Lond.* 281, 40–46.

Martin, R. G. (1981). The transformation of cell growth and transmogrification of DNA synthesis by Simian virus 40. *Adv. Cancer Res.* 34, 1–68.

Mathis, D. J. and Chambon, P. (1981). The SV40 early region TATA box is required for accurate *in vitro* initiation of transcription. *Nature, Lond.* 290, 310–315.

Maxam, A. M. and Gilbert, W. (1980). Sequencing end-labeled DNA with base-specific chemical cleavages. *In* "Methods in Enzymology" (Eds L. Grossman and K. Moldave) Vol 65, 499—560. Academic Press, New York.

May, E., Kress, M., Daya-Grosjean, L., Monier, R. and May, P. (1981). Mapping of the viral mRNA encoding a super T-antigen of 115,000 daltons expressed in simian virus 40-transformed rat cell lines. *J. Virol.* **37**, 24—35.

Mellon, P., Parker, V., Gluzman, Y. and Maniatis, T. (1981). Identification of DNA sequences required for transcription of the human α_1-globin gene in a new SV40 host-vector system. *Cell* **27**, 279—288.

Moar, M. H., Campo, M. S., Laird, H. M. and Jarrett, W. F. H. (1981a). Unintegrated viral DNA sequences in a hamster tumour induced by bovine papilloma virus. *J. Virol.* **39**, 945—949.

Moar, M. H., Campo, M. S., Laird, H. and Jarrett, W. F. H. (1981b). Persistence of non-integrated viral DNA in bovine cells transformed *in vitro* by bovine papillomavirus type 2. *Nature, Lond.* **293**, 749—751.

Moriarty, A. M., Hoyer, B. H., Shih, J. W-K., Gerin, J. L. and Hamer, D. H. (1981). Expression of the hepatitis B virus surface antigen gene in cell culture by using a Simian virus 40 vector. *Proc. Natn. Acad. Sci. U.S.A.* **78**, 2606—2610.

Moschonas, N., de Boer, E., Grosveld, F. G., Dahl, H. H. M., Wright, S., Shewmaker, C. K. and Flavell, R. A. (1981). Structure and expression of a cloned β^0 thalassaemic globin gene. *Nucl. Acids Res.* **9**, 4391—4401.

Mulligan, R. C. and Berg, P. (1980). Expression of a bacterial gene in mammalian cells. *Science, N.Y.* **209**, 1422—1427.

Mulligan, R. C. and Berg, P. (1981a). Selection for animal cells that express the *Escherichia coli* gene coding for xanthine—guanine phosphoribosyltransferase. *Proc. Natn. Acad. Sci. U.S.A.* **78**, 2072—2076.

Mulligan, R. C. and Berg, P. (1981b). Factors governing the expression of a bacterial gene in mammalian cells. *Molec. Cell. Biol.* **1**, 449—459.

Mulligan, R. C., Howard, B. H. and Berg, P. (1979). Synthesis of rabbit β-globin in cultured monkey kidney cells following infection with a SV40-β-globin recombinant genome. *Nature, Lond.* **277**, 108—114.

Mulligan, R. C., White, R. T. and Berg, P. (1980). Formation of β-globin following infection with recombinants containing rabbit β-globin cDNA at different locations of SV40's late region. *In* "Mobilisation and Reassembly of Genetic Information" (Eds W. A. Scott and R. Werner). Miami Winter Symposia, Vol. 17, 201—215. Academic Press, New York.

Mulligan, R. C., Southern, P. J., Howard, B. H., Yaniv, M., Geller, A. I. and Berg, P. (1982). Construction and potential uses for a family of mammalian transducing vectors. *J. Molec. Appl. Genet.*, in press.

Muzyczka, N. (1979). Persistence of phage λ DNA in genomes of mouse cells transformed by λ-carrying SV40 vectors. *Gene* **6**, 107—122.

Myers, R. M. and Tjian, R. (1980). Construction and analysis of simian virus 40 origins defective in tumor antigen binding and DNA replication. *Proc. Natn. Acad. Sci. U.S.A.* **77**, 6491—6495.

Nunberg, J. H., Kaufman, R. J., Chang, A. C. Y., Cohen, S. N. and Schimke, R. T. (1980). Structure and genomic organisation of the mouse dihydrofolate reductase gene. *Cell* **19**, 355—364.

O'Hare, K., Benoist, C. and Breathnach, R. (1981). Transformation of mouse fibroblasts to methotrexate resistance. *Proc. Natn. Acad. Sci. U.S.A.* **78**, 1527—1531.

Proudfoot, N. J. and Brownlee, G. G. (1976). 3' non-coding region sequences in eukaryotic messenger RNA. *Nature, Lond.* **263**, 211—214.

Rassoulzadegan, M., Binetruy, B. and Cuzin, F. (1982). High frequency of gene transfer after fusion between bacteria and eukaryotic cells. *Nature, Lond.* **295**, 257—259.

Rigby, P. W. J., Chia, W., Clayton, C. E. and Lovett, M. (1980). The structure and expression of the integrated viral DNA in mouse cells transformed by Simian virus 40. *Proc. R. Soc. Lond. B* **210**, 437—450.

Rio, D., Robbins, A., Myers, R. and Tjian, R. (1980). Regulation of Simian virus 40 early transcription *in vitro* by a purified tumor antigen. *Proc. Natn. Acad. Sci. U.S.A.* **77**, 5706—5710.

Sager, R., Anisowicz, A. and Howell, N. (1981). Genomic rearrangements in a mouse cell line containing integrated SV40 DNA. *Cell* **23**, 41—50.

Sambrook, J. and Grodzicker, T. (1980). Adenovirus-SV40 hybrids: a model system for expression of foreign sequences in an animal virus vector. *In* "Genetic Engineering: Principles and Methods" (Eds J. K. Setlow and A. Hollaender) Vol. 2, 103—114. Plenum Press, New York and London.

Sanger, F. (1981). Determination of nucleotide sequences in DNA. *Bioscience Rep.* **1**, 3—18.

Sarver, N., Gruss, P., Law, M-F., Khoury, G. and Howley, P. M. (1981a). Bovine papilloma virus deoxyribonucleic acid: a novel eucaryotic cloning vector. *Molec. Cell. Biol.* **1**, 486—496.

Sarver, N., Gruss, P., Law, M-F., Khoury, G. and Howley, P. M. (1981b). Rat insulin gene covalently linked to bovine papillomavirus DNA is expressed in transformed mouse cells. *In* "Developmental Biology Using Purified Genes" (Eds D. Brown and C. R. Fox). ICN-ULCA Symposia on Molecular and Cellular Biology, Vol. 23. Academic Press, New York. In press.

Scangos, G. and Ruddle, F. H. (1981). Mechanisms and applications of DNA-mediated gene transfer in mammalian cells; a review. *Gene* **14**, 1—10.

Schaffner, W. (1980). Direct transfer of cloned genes from bacteria to mammalian cells. *Proc. Natn. Acad. Sci. U.S.A.* **77**, 2163—2167.

Schaffner W., Topp, W. and Botchan, M. (1979). Simian virus 40 used as a transforming vector for the insertion of foreign DNA into rat cell chromosomes. *Experientia* **35**, 977.

Schimke, R. T., Brown, P. C., Kaufman, R. F., McGrogan, M. and Slate, D. L. (1981) Chromosomal and extrachromosomal localization of amplified dihydrofolate reductase genes in cultured mammalian cells. *Cold Spring Harb. Symp. Quant. Biol.* **45**, 785—798.

Shortle, D. and Nathans, D. (1978). Local mutagenesis: a method for generating viral mutants with base substitutions in preselected regions of the viral genome. *Proc. Natn. Acad. Sci. U.S.A.* **75**, 2170—2174.

Soberon, X., Covarrubias, L. and Bolivar, F. (1980). Construction and characterization of new cloning vehicles. IV. Deletion derivatives of pBR322 and pBR325. *Gene* **9**, 287—305.

Solnick, D. (1981). Construction of an adenovirus-SV40 recombinant producing SV40 T-antigen from an adenovirus late promoter. *Cell* **24**, 135—143.

Southern, E. M. (1975). Detection of specific sequences among DNA fragments separated by agarose gel electrophoresis. *J. Molec. Biol.* **98**, 503—515.

Stow, N. (1981). Cloning of a DNA fragment from the left-hand terminus of the adenovirus type 2 genome and its use in site-directed mutagenesis. *J. Virol.* **37**, 171—180.

Subramani, S., Mulligan, R. and Berg, P. (1981). Expression of the mouse dihydrofolate reductase complementary deoxyribonucleic acid in Simian virus 40 vectors. *Molec. Cell. Biol.* **1**, 854—864.

Sveda, M. M. and Lai, C-J. (1981). Functional expression in primate cells of cloned DNA coding for the hemagglutinin surface glycoprotein of influenza virus. *Proc. Natn. Acad. Sci. U.S.A.* **78**, 5488—5492.

Thomas, P. S. (1980). Hybridization of denatured RNA and small DNA fragments transferred to nitrocellulose. *Proc. Natn. Acad. Sci. U.S.A.* **77**, 5201—5205.

Thummel, C., Tjian, R. and Grodzicker, T. (1981). Expression of SV40 T antigen under control of adenovirus promoters. *Cell* **23**, 825—836.

Tooze, J. (1980). "Molecular Biology of Tumor Viruses". Part 2, DNA Tumor Viruses. 2nd edn. Cold Spring Harbor Laboratory, New York.

Topp, W. C. (1981). Normal rat cell lines deficient in nuclear thymidine kinase. *Virology* **113**, 408—411.

Treisman, R., Novak, U., Favaloro, J. and Kamen, R. (1981). Transformation of rat cells by an altered polyoma virus genome expressing only the middle-T protein. *Nature, Lond.* **292**, 595—600.

Twigg, A. J. and Sherratt, D. (1980). Trans-complementable copy-number mutants of plasmid ColE1. *Nature, Lond.* **283**, 216—218.

Upcroft, P., Skolnik, H., Upcroft, J. A., Solomon, D., Khoury, G., Hamer, D. H. and Fareed, G. C. (1978). Transduction of a bacterial gene into mammalian cells. *Proc. Natn. Acad. Sci. U.S.A.* **75**, 2117—2121.

Villarreal, L. P. and Berg, P. (1977). Hybridization *in situ* of SV40 plaques: Detection of recombinant SV40 virus carrying specific sequences of nonviral DNA. *Science, N.Y.* **196**, 183—185.

Wagner, T. E., Hoppe, P. C., Jollick, J. D., Scholl, D. R., Hodinka, R. L. and Gault, J. B. (1981). Microinjection of a rabbit β-globin gene into zygotes and its subsequent expression in adult mice and their offspring. *Proc. Natn. Acad. Sci. U.S.A.* **78**, 6376—6380.

Wallace, R. B., Shaffer, J., Murphy, R. F., Bonner, J., Hirose, T. and Itakura, K. (1979). Hybridization of synthetic oligodeoxyribonucleotides to ØX174 DNA: the effect of single base pair mismatch. *Nucl. Acids Res.* **6**, 3543—3557.

Weber, H., Dierks, P., Meyer, F., van Ooyen, A., Dobkin, C., Abrescia, P., Kappeler, M., Mayhack, B., Zeltner, A., Mullen, E. E. and Weissmann, C. (1981). Modification of the rabbit chromosomal β-globin gene by restructuring and site-directed mutagenesis. *In* "Developmental Biology Using Purified Genes" (Eds D. Brown and C. R. Fox). ICN-UCLA Symposia on Molecular and Cellular Biology, Vol. 23,. Academic Press, New York. In press.

Wei, C-M., Gibson, M., Spear, P. G. and Scolnick, E. M. (1981). Construction and isolation of a transmissible retrovirus containing the *src* gene of Harvey murine sarcoma virus and the thymidine kinase gene of herpes simplex virus type 1. *J. Virol.* **39**, 935—944.

Weil, P. A., Luse, D. S., Segall, J. and Roeder, R. G. (1979). Selective and accurate initiation of transcription at the Ad2 major late promoter in a soluble system dependent on purified RNA polymerase II and DNA. *Cell* **18**, 469—484.

Weingartner, B. and Keller, W. (1981). Transcription and processing of adenoviral RNA by extracts from HeLa cells. *Proc. Natn. Acad. Sci. U.S.A.* **78**, 4092—4096.

Weiss, R. A., Teich, N. M., Varmus, H. and Coffin, J. M. (1982). "Molecular Biology of Tumor Viruses". Part 3, RNA Tumor Viruses. 2nd edn. Cold Spring Harbor Laboratory, New York. In press.

Weissmann, C., Weber, H., Taniguchi, T., Muller, W. and Meyer, F. (1979). Application of site-directed mutagenesis to ribonucleic acid and deoxyribonucleic acid genomes. *In* "Biochemistry of Genetic Engineering" (Eds P. B. Garland and R. Williamson). *Biochem. Soc. Symp.* **44**, 43—55.

Wigler, M., Sweet, R., Sim, G. K., Wold, B., Pellicer, A., Lacy, E., Maniatis, T., Silverstein, S. and Axel, R. (1979). Transformation of mammalian cells with genes from procaryotes and eukaryotes. *Cell* **16**, 777—785.

Wozney, J., Hanahan, D., Tate, V., Boedtker, H. and Doty, P. (1981). Structure of the proα2(I) collagen gene. *Nature, Lond.* **294**, 129—135.

A comprehensive list of cloned eukaryotic genes

KAY E. DAVIES

Biochemistry Department, St Mary's Hospital
Medical School, University of London, London, UK

The following list of references includes those describing most of the genomic and cDNA recombinant molecules corresponding to eukaryotic coding sequences that have been cloned in various laboratories since the first recombinant molecule was constructed in the early 1970s. It is hoped that those entering this rapidly advancing field will be able to benefit from this list because it provides a quick index to structural gene clones already available. It should also be useful for those cloning new genes as many types of proteins are included, and the miscellaneous sections indicate more general experimental approaches.

Not all of the available literature has been cited and characterizations of clones already described have, in general, been omitted. If a particular gene has been cloned in several laboratories, then each of the important references are given so that the reader will be able to follow the work in detail. One or two prokaryotic genes that are of potential interest for the transformation of eukaryotic organisms are included, such as the nitrogen fixation genes.

Inevitably, some genes will have been missed and new ones cloned since this article was written (September, 1981). Additional references will be added in future volumes of *Genetic Engineering* and I hope that researchers in the field will communicate their publications to me for inclusion.

Key

Table 1 gives an index by number of all the clones listed according to the corresponding gene product.

Table 2 gives an index by number of all the clones listed according to their origin. Plant genes are grouped together.

The main list is divided into two sections, one for genomic sequences and the other for cDNA sequences. Within this classification the genes are listed alphabetically according to the corresponding gene product. A miscellaneous section is included at the end of each section for cloned structural genes whose products are not defined.

Table 1

| | Reference number | |
| | Genomic clone | cDNA clone |
Gene product		
ACTH β-LPH	1	
Actin	2, 3, 4, 5, 6	179, 180, 181, 182, 183, 184, 185, 283
Albumin	7, 8	186, 187, 188
Alcohol dehydrogenase	9, 10	
Alpha-fetoprotein	11, 12, 13	188, 189
Amylase	14, 15	
Antifreeze peptide		190
Apoprotein		191
Aprt (adenine phosphoribosyl transferase)	16	
Arginosuccinate lyase (Arg H)	17	
Calmodulin		192
Casein		193
CDC10 gene	18	
Cell cycle gene	19	
Chorion	20, 21	194, 195, 196, 197
Collagen	22, 23, 24	198, 199, 200, 201
Complement (C3)		202
Conalbumin	25, 26	203
Corticotropin		204
Crystallin	27	205
Cuticle genes	28	
Cytochrome c	29, 30, 31, 32	
Cytochrome oxidase	33	
Dehydroquinase	34	
Dihydrofolate reductase	35	207
Discoidin-1	36	
β-Endorphin		204
Fibroin	38, 39	
Foot and mouth virus antigen		208

Table 1 continued

Gene product	Reference number	
	Genomic clone	cDNA clone
Galactokinase	40	
β-Galactosidase	41	
Globin	42, 43, 44, 45, 46, 47, 48, 49, 50, 51	209, 210, 211, 212, 213, 214, 215, 216, 217, 218
α2μ-globulin		219, 220, 221
Glycero-3-phosphate dehydrogenase	52	
Gonadotropin		222, 223
Growth hormone	53, 54, 124	224, 225, 226, 227
Heat shock protein	57, 58, 59, 60, 61	229
His 3	63, 80	
His B	64	
Histocompatibility antigen	65	230, 231, 232
Histone	66, 67, 68, 69, 70, 71, 72, 73, 74, 75, 76, 77, 78, 79, 132	
Imidazole-glycerol phosphate dehydratase	63, 80	
Immunoglobulin	81, 82, 83, 84, 85, 86, 87, 88, 89, 90, 91, 92, 93, 94	233, 234, 235, 236, 237, 238, 239
Insulin	95, 96, 97, 98	240, 241, 242, 243
Interferon	99, 100, 101	244, 245, 246, 247, 248, 249, 250
Intracisternal A particle genes	102	
Lactalbumin		193, 251
Larval serum protein		252
Leghaemoglobin	103, 104	253
Leu 2	105	
Lipoprotein		254
Lysozyme	106, 107	255, 262
Metallothionen-I	108	
β-2 microglobulin		256
Mitochondrial genes	109, 110, 111, 112, 113	
Milk proteins		257
Myosin	114, 115	179, 258, 259
Nitrogen fixation genes	116, 117	

Table 1 continued

Gene product	Reference number	
	Genomic clone	cDNA clone
Ovalbumin	118, 119	260, 261
Ovomucoid	120	262, 263
Parathyroid hormone		264, 265
Phaseolin	121	
3-Phospholglycerokinase (PGK)	122	
Prolactin	123	266, 267, 268
Pro-opiomelanocortin peptide	124	
Protamine		269, 270
Relaxin		271
Ribosomal protein	128, 129	272, 273
Ribosomal RNA, 5 S	130, 131, 132, 133, 134, 135, 136	
Ribosomal RNA	137, 138, 139, 140, 141, 142, 143, 144, 145, 146	
Ribulose-1,5-biphosphate carboxylase	147, 148	
Seminal vesicle secretory proteins IV and V		274
Somatomammotropin		275, 276
Sucrose synthetase		277
Thrombin		279
Thymidine kinase	149, 150	
Thyroglobulin		280
Transfer RNA	151, 152, 154, 155, 156, 157, 158, 159, 160, 161	
Transfer RNA synthetase	153	
Transferrin	162	
Transplantation genes		281
α-Tropomysin		282
Tubulin	164	283, 284
Uteroglobulin		285
U5 RNA	165	
U1 RNA	166	
Ura-1	167	
Viral genes	55, 56, 62, 176, 178	228
Vitellogenin	168, 169, 170, 171	286, 287, 288, 289
Yolk proteins	172	

Table 2

Origin	Reference number
Carp	242
Cow	227, 264, 265, 268, 279, 280
Chicken	22, 24, 25, 26, 27, 70, 71, 98, 106, 107, 118, 119, 120, 149, 161, 162, 165, 166, 183, 184, 186, 191, 198, 199, 201, 203, 205, 209, 215, 254, 255, 258, 259, 260, 261, 262, 263, 282, 283, 286, 287, 290, 294
Dictyostelium discoideum	36, 138, 180, 292, 296
Drosophila melanogaster	2, 6, 9, 28, 37, 57, 58, 59, 60, 61, 74, 130, 133, 136, 140, 141, 143, 145, 151, 152, 154, 155, 164, 168, 172, 174, 175, 194, 196, 229, 252, 284, 290
Eel	192
Hamster	16, 207
Human	42, 44, 46, 49, 50, 51, 53, 67, 72, 88, 95, 99, 100, 101, 109, 111, 124, 165, 166, 200, 213, 217, 222, 223, 224, 225, 232, 240, 244, 245, 246, 247, 248, 249, 250, 251, 266, 275, 276
Mouse	11, 12, 13, 35, 43, 45, 65, 81, 82, 83, 84, 85, 86, 87, 89, 90, 91, 92, 93, 94, 102, 108, 110, 128, 137, 182, 188, 202, 204, 206, 210, 230, 231, 233, 235, 236, 237, 238, 239, 256, 273, 281
Neurospora crassa	34
Nematode	114
Newts	76
Plants	103, 104, 121, 139, 142, 147, 148, 253, 277
Rabbit	47, 48, 214, 216, 218, 277, 285
Rat	1, 7, 8, 14, 30, 54, 96, 97, 115, 123, 159, 160, 165, 173, 179, 185, 187, 189, 193, 219, 220, 221, 226, 241, 243, 257, 267, 271, 274, 293, 295
Sea urchin	5, 66, 68, 69, 73, 77, 181
Sheep	23
Silk moth	20, 21, 38, 39, 195, 197
Tetrahymena pyroformis	146
Trout	269, 270
Viruses	55, 56, 62, 150, 173, 176, 178, 208, 228
Winter flounder	190
Xenopus laevis	75, 79, 127, 131, 132, 134, 135, 169, 170, 171, 211, 212, 234, 272, 288, 289, 291
Yeast	3, 4, 10, 17, 18, 19, 29, 31, 32, 33, 40, 41, 52, 63, 64, 78, 80, 105, 112, 113, 122, 125, 126, 129, 144, 153, 156, 157, 158, 163, 167, 177

Genomic clones

A

1. Drouin, J. and Goodman, H. M. (1980) *Nature* **288**: 610. Most of the coding region of rat ACTH-β-LPH precursor gene lacks intervening sequences.
2. Fyrberg, E. A., Kindle, K. L., Davidson, N. and Sodja, A. (1980) *Cell* **19**: 365. The actin genes of *Drosophila*: a dispersed multigene family.
3. Gallwitz, D. and Seidel, R. (1980) *Nucl. Acids Res.* **8**: 1043 Molecular cloning of the actin genes from yeast *Saccharomyces cerevisiae*.
4. Ny, R. and Abelson, J. (1980) *Proc. Nat. Acad. Sci. USA* **77**: 3912. Isolation and sequence of the gene for actin in *Saccharomyces cerevisiae*.
5. Schuler, M. A. and Keller, E. B. (1981) *Nucl. Acids Res.* **9**: 591. The chromosomal arrangement of two linked actin genes in the sea urchin *S. purpuratus*.
6. Tobin, S. L., Zulauf, E. Sanchez, F., Craig, E. A. and McCarthy, B. J. (1980) *Cell* **19**: 121. Multiple actin-related sequences in the *Drosophila melanogaster* genome.
7. Kiossis, D., Hamilton, R., Hanson, R. W., Tilghman, S. M. and Taylor, J. M. (1979) *Proc. Nat. Acad. Sci. USA* **76**: 4370. Construction and cloning rat albumin structural gene sequences.
8. Sargent, T. D., Wu, J-R., Sala-Trepat, J-M., Wallace, R. B., Reyes, A. A. and Bonner, J. (1979) *Proc. Nat. Acad. Sci. USA* **76**: 3256. The rat serum albumin gene: analysis of cloned sequences.
9. Benyajati, C., Place, A. R., Powers, D. A. and Sofer, W. (1981) *Proc. Nat. Acad. Sci. USA* **78**: 2717. Alcohol dehydrogenase gene of *Drosophila melanogaster* — relationship of intervening sequences to functional domains in the protein.
10. Williamson, V. M., Bennetzen, J., Young, E. T., Nasmyth, K. and Hall, B. D. (1980) *Nature* **283**: 214. Isolation of the structural gene for alcohol dehydrogenase by genetic complementation in yeast.
11. Gorin, M. B. and Tilghman, S. M. (1980) *Proc. Nat. Acad. Sci. U.S.A* **77**: 1351. Structure of the alpha-fetoprotein gene in the mouse.
12. Kiossis, D., Eiferman, F., van de Rijn, P., Gorin, M. B., Ingram, R. S. and Tilghman, S. M. (1980) *J. Biol. Chem.* **256**: 1960. The evolution of α-fetoprotein and albumin genes II. The structure of α-fetoprotein and albumin in the mouse.
13. Tilghman, S. M., Kiossis, D., Gorin, M. B. Ruiz, J. P. G. and

Ingram, R. S. (1979) *J. Biol. Chem.* 254: 7393. The presence of intervening sequences in the α-fetoprotein gene of the mouse.

14. MacDonald, R. J., Grerar, M. M., Swain, W. F., Pictet, R. L., Thomas, G. and Rutter, W. J. (1980) *Nature* 287. 117. Structure of a family of rat amylase genes.

15. Young, R. A., Hagenbuehle, O. and Schibler, U. (1981) *Cell* 23: 451. A single mouse α-amylase gene specifies two different tissue specific mRNAs.

16. Lowy, I., Pellicer, A., Jackson, J. F., Sim, G-K, Silverstein, S. and Axel, R. (1980) *Cell* 22: 817. Isolation of transforming DNA: cloning the hamster aprt gene.

17. Clarke, L. and Carbon, J. (1978) *J. Mol. Biol.* 120: 517. Functional expression of cloned yeast DNA in *E. Coli*: specific complementation of arginosuccinate lyase (arg H) mutations.

C

18. Clarke, L. and Carbon, J. (1980) *Proc. Nat. Acad. Sci. USA* 77: 2173. Isolation of the centromere linked CDC10 gene by complementation in yeast.

19. Nasmyth, K. A. and Reed, S. I. (1980) *Proc. Nat. Acad. Sci. USA* 77: 2119. Isolation of genes by complementation in yeast — molecular cloning of a cell-cycle gene.

20. Jones, C. W. and Kafatos, F. C. (1980) *Cell* 22: 855. Structure, organisation and evolution of developmentally regulated chorion genes in a silkmoth.

21. Weldon-Jones, C. and Kafatos, F. C. (1980) *Nature* 284: 635. Co-ordinately expressed members of two chorion multigene families are clustered, alternating and divergently orientated.

22. Ohkubo, H., Vogeli, G., Mudryi, M., Avvedimento, V. E., Sullivan, M., Paston, I. and DeCrombrugghe, B. (1980) *Proc. Nat. Acad. Sci. USA* 77: 7059. Isolation and characterisation of overlapping genomic clones covering the chicken alpha-2 (Type 1) collagen gene.

23. Schafer, M. P., Boyd, C. D., Tolstoshev, P. and Crystal, R. G. (1980) *Nucl. Acids Res.* 8: 2241. Structural organisation of a 17 kb segment of the α2 collagen gene: evaluation by R loop mapping.

24. Vooeli, G., Avvedimento, E. V., Sullivan, M. A., Maizel, J. V., Lozano, G., Adams, S. L., Paston, I. and DeCrombrugghe, B. (1980) *Nucl. Acids Res.* 8: 1823. Isolation and characterisation of genomic DNA cloning for α2 Type 1 collagen.

25. Cochet, M., Ganna, F., Hen, R., Maroteaux, L., Perrin, F. and Chambon, P. (1979) *Nature* 282: 567. Organisation and sequence studies of the 17-piece chicken conalbumin gene.

26. Perrin, F., Cochet, M., Gerlinger, P., Cami, B., Lepennec, J. P. and Chambon, P. (1979) *Nucl. Acids Res.* 6: 2731. The chicken conalbumin gene: studies of the organisation of cloned DNAs.
27. Bhat, S. P., Jones, R. E., Sullivan, M. A. and Piatigorsky, J. (1980) *Nature* 284: 234. Chicken lens crystallin DNA sequences show at least two δ- crystallin genes.
28. Synder, M., Hirsh, J. and Davidson, N. (1981) *Cell* 25: 165. The cuticle genes of *Drosophila*: a developmentally regulated gene cluster.
29. Montgomery, D. L., Hall, B. D., Gillan, S. and Smith, M. (1978) *Cell* 4: 673. Identification and isolation of the yeast cytochrome c gene.
30. Scarpulla, R. C., Agne, K. M. and Wu, R. (1981) *J. Biol. Chem.* 256: 6480. Isolation and structure of rat cytochrome c gene.
31. Montgomery, D. L., Leung, D. W., Smith, M., Shalit, P., Faye, G. and Hall, B. D. (1980) *Proc. Nat. Acad. Sci. USA* 77: 541. Isolation and sequence of the gene for iso-2-cytochrome c in *Saccharomyces-cerevisiae*.
32. Szostak, J. W., Stiles, J. I., Tye, B-K., Chin, P., Sherman, F. and Wu, R. (1979) *Methods in Enzymology* 68: 419. Hybridisation with synthetic oligonucleotides.
33. Bonitz, S. G., Corruzzi, G., Thanfeld, B. E., Tzagoloff, A. and Macino, G. (1980) *J. Biol. Chem.* 255: 11927. Assembly of the mitochondrial membrane system. Structure and nucleotide sequence of the gene coding for subunit 1 of yeast cytochrome oxidase.

D

34. Alton, N. D., Hautala, J. A., Giles, N. H., Kushner, S. R. and Vapnek, D. (1978) *Gene* 4: 241. Transcription and translation in *E. coli* of hybrid plasmids containing the catabolic dehydroquinase gene from *Neurospora crassa*.
35. Nunberg, J. H., Kaufman, R. J., Chang, A. C. Y., Cohen, S. N. and Schimke, R. T. (1980) *Cell* 19: 355. Structure and genomic organisation of the mouse dihdrofolate reductase gene.
36. Rowenkamp, W., Poole, S. and Firtel, R. A. (1980) *Cell* 20: 495. Analysis of the multigene family coding the developmentally regulated carbohydrate-binding protein Discoidin-1 in *D. discoideum*.

E

37. Vilek, M. P., Kreissman, S. G. and Gross, R. H. (1981) *Nucl.*

Acids Res. 9: 1191. The isolation of ecdysterone inducible genes by hybridisation subtraction chromatography

F
38. Oshima, Y. and Suzuki, Y. (1977) *Proc. Nat. Acad. Sci. USA* 74: 5363. Cloning of the silk fibroin gene and its flanking sequences.
39. Tsujimoto, Y. and Suzuki, Y. (1975) *Cell* 18: 591. The DNA sequence of *Bambyx mori* fibroin gene including the 5′ flanking, mRNA coding, entire intervening and fibroin protein coding regions.

G
40. Schell, M. A. and Wilson, D. B. (1979) *Gene* 5: 291. Cloning and expression of the yeast galactokinase gene in an *E. coli* plasmid.
41. Markeri, J. G. and Dickson, R. C. (1978) *Cell* 15: 123. Molecular cloning and expression in *E. coli* of yeast gene coding for beta-galactosidase.
42. Lauer, J., Shen, C.-K. J. and Maniatis, T. (1980) *Cell* 20: 119. The chromosomal arrangement of human alpha-like globin genes: sequence homology and alpha-globin gene deletions.
43. Nishioka, Y., Ledo, A. and Leder, P. (1980) *Proc. Nat. Acad. Sci. USA* 77: 2806. Unusual alpha-globin-like gene that has clearly lost both globin intervening sequences.
44. Ramirez, F., Burns, A. L., Mears, J. G., Spence, S., Starkman, D. and Bank, A. (1979) *Nucl. Acids Res.* 7: 1147. Isolation and characterisation of cloned human fetal globin genes.
45. Vanin, E. F., Goldberg, G. I., Tucker, P. W. and Smithies, O. (1980) *Nature* 286: 232. A mouse alpha-globin-related pseudo-gene lacking intervening sequences.
46. Fritsch, E. F., Lawn, R. M. and Maniatis, T. (1980) *Cell* 19: 959. Molecular cloning and characterisation of the human beta-like globin gene cluster.
47. Lacy, E., Fahrner, K., Woolford, J. Jr., Rosbash, M. and Kaback, D. B. (1979) *Cell* 18: 1273. The linkage arrangement of four rabbit beta-like globin genes.
48. Jeffreys, A. J. and Flavell, R. A. (1977) *Cell* 12: 1097. The rabbit beta-globin gene contains a large insert in the coding sequence.
49. Bernards, R., Little, P. F. R., Annison, G., Williamson, R. and Flavell, R. A. (1979) *Proc. Nat. Acad. Sci. USA* 79: 4827. Structure of the human G-gamma-A-gamma-delta-beta-globin gene locus.

50. Blattner, F. R., Blechl, A. E., Denniston-Thompson, K., Faber, H. E., Richards, J. E., Slightom, J. L., Tucker, P. W. and Smithies, O. (1978) *Science* **202:** 1279. Cloning of human fetal gamma globin and mouse alpha-type globin DNA: preparation and screening of shotgun collections.

51. Proudfoot, N. J. and Baralle, F. E. (1979) *Proc. Nat. Acad. Sci. USA* **76:** 5435. Molecular cloning of human epsilon-globin gene.

52. Holland, J. P. and Holland, J. (1979). *J. Biol. Chem* **254:** 5466. Isolation and characterisation of a gene coding for glyceraldehyde-3-phosphate dehydrogenase from *Saccharomyces cerevisiae*.

53. Fiddes, J.C., Seeburg, P. H., Denoto, F. M., Hallewell, R. A., Baxter, J. D. and Goodman, H. M. (1979) *Proc. Nat. Acad. Sci. USA* **76:** 4294. Structure of genes for human growth hormone and chorionic somatomammotropin.

54. Page, G. S., Smith, S. and Goodman, H. M. (1981) *Nucl. Acids Res.* **9:** 2087. DNA sequence of rat growth hormone gene: location of the 5′ terminus of the growth hormone messenger RNA and identification of an internal transposon-like element.

H

55. Both, G. W. and Sleigh, M. J. (1980) *Nucl. Acids Res* **8:** 2561. Complete nucleotide sequence of the haemagglutinin gene from a human influenza virus of the Hag Keng subtype.

56. Gething, M. J., Bye, J., Skehel, J. and Waterfield, M. (1980) *Nature* **287:** 301. Cloning and DNA sequence of double stranded copies of haemagglutin genes from H2 and H3 strains elucidates antigenic shift and drift in human influenza virus.

57. Craig, E. A. and McCarthy, B. J. (1980) *Nucl. Acids Res.* **8:** 4441. Four *Drosophila* heat shock genes at 67B: characterisation of recombinant plasmids.

58. Goldschmidt-Clermont, M. (1980) *Nucl. Acids Res* **8:** 235. Two genes for the major-heat-shock protein of *Drosophila melanogaster*.

59. Holmgren, R., Livak, K., Morimoto, R., Freund, R. and Moselvn, M. (1979) *Cell* **18:** 1359. Studies of cloned sequences from four *Drosophila* heat shock loci.

60. Mirault, M. E., Goldschmidt-Clermont, M., Artavanistsakonas, S. and Scheal, P. (1979) *Proc. Nat. Acad. Sci. USA* **76:** 5254. Organisation of the multiple genes for the 70,000 dalton heat shock protein in *Drosophila melanogaster*.

61. Wadsworth, S. C., Craig, E. A. and McCarthy, B. J. (1980) *Proc. Nat. Acad. Sci. USA*, **77,** 2134. Genes for three *Drosophila* heat-shock-induced proteins at a single locus.

62. Galibert, F., Mardart, E., Fitoussi, F., Tiollais, P. and Charnay, C. (1979) *Nature* 281: 646. Nucleotide sequence of the *hepatitis B virus* genome (subtype ayw) cloned in *E. coli*.
63. Davis, R. W. and Struhl, K. (1980) *J. Mol. Biol.*, **136**, 309. Genetic and transcriptional map of the cloned his 3 region of *Saccharomyces cerevisiae*.
64. Struhl, K. and Davis, R. W. (1976) *Proc. Nat. Acad. Sci. USA* 73: 1471. Functional expression of eukaryotic DNA in *E. coli* (his B).
65. Cami, B., Bregegere, F., Abastado, J. P. and Kourilsky, P. (1981) *Nature* 291: 673. Multiple sequences related to classical histocompatibility antigens in the mouse genome.
66. Birnsteil, M. L., Schaffner, W. and Smith, H. O. (1977) *Nature* 266: 603. DNA sequences coding for the H2B histone of *Psammechinus miliaris*.
67. Clark, S. J., Krieg, P. A. and Wells, J. R. E. (1981) *Nucl. Acids Res.* 9: 1583. Isolation of a clone containing human histone genes.
68. Cohn, R. H. and Kedes, L. H. (1979) *Cell* 18: 843. Nonallelic histone gene clusters of individual sea urchins (*Lytechinus pictus*).
69. Cohn, R. H., Lowry, J. C. and Kedes, L. H. (1976) *Cell* 9: 147. Histone genes of the sea urchin cloned in *E. coli*: order, polarity and strandedness of five histone-coding and spacer regions.
70. Engel, J. D., and Dodgson, J. B. (1981) *Proc. Nat. Acad. Sci. USA* 78: 2856. Histone genes are clustered but not tandemly repeated in the chicken genome.
71. Harvey, R. P. and Wells, J. R. E. (1979) *Nucl. Acids Res.* 7: 1787. Isolation of a genomic clone containing chicken histone genes.
72. Heintz, N., Zernik, M. and Roeder, R. G. (1981) *Cell* 24: 661. The structure of the human histone genes: clustered but not tandemly repeated.
73. Kedes, L. H., Chang, A. C. Y., Housman, D. and Cohen, S. N. (1975) *Nature* 255: 533. Isolation of histone genes from unfractionated sea urchin DNA by subculture cloning in *E. coli*.
74. Lifton, R. P., Goldberg, M. L., Karo, R. W. and Hogness, D. S. (1977) *Cold Spring Harbour Symp. Quant. Biol.* 42: 1047. The organisation of the histone genes in *Drosophila melanogaster*: functional and evolutionary implications.
75. Moorman, A. F. M., Delaaf, R. T. M., Destree, O. H. J., Telford, J. and Birnstiel, M. L. (1980) *Gene* 10: 185. Histone genes from *Xenopus laevis*-molecular cloning and initial characterisation.

76. Stephenson, E. C., Erba, H. P. and Gall, J. G. (1981) *Cell* 24: 639. Histone gene clusters of the newt *Notophthalmus* are separated by long tracts of satellite DNA.
77. Sures, I., Maxam, A., Cohn, R. H. and Kedes, L. H. (1976) *Cell* 9: 495. Identification and location of the histone H2A and H3 genes by sequence analysis of sea urchin (*S. purpuratus*) DNA cloned in *E. coli.*
78. Hereford, L. M., Fahrner, K., Woolford, J. L. Jr. and Rosbash, M. (1979) *Cell* 18: 1261. Isolation of yeast histone genes H2A and H2B.
79. Zernik, M., Heintz, N. Boime, I. and Roeder, R. G. (1980) *Cell* 22: 807. *Xenopus laevis* histone genes: variant H1 genes are present in different clusters.

I

80. Struhl, K. and Davis, R. W. (1977) *Proc. Nat. Acad. Sci. USA* 74: 5255. Production of a functional eukaryotic enzyme in *E. coli*: cloning and expression of the yeast structural gene for imidazole-glycerolphosphate dehydratase (his 3).
81. Adams, J. M., Webb, E., Gerondakis, S. and Cory, S. (1980) *Nucl. Acids Res.* 8: 6019. Cloned embryonic DNA sequences flanking the mouse immunoglobulin C-γ-3 and C-γ-1 genes.
82. Auffray, C., Nageotte, R., Chambraud, B. and Rougeon, R. (1980) *Nucl. Acids. Res.* 8: 1231. Mouse immunoglobulin genes: a bacterial plasmid containing the entire coding sequence for a pre-γ-2a Leary chain.
83. Blomberg, B., Zraunecker, A., Eisen, H. and Tonegawa, S. (1981) *Proc. Nat. Acad. Sci. USA* 78: 3765. Organisation of four mouse λ light chain immunoglobulin genes.
84. Calane, K., Rogers, J., Early, P. Davis, H., Livant, D., Wall, R. and Hood, L. (1980) *Nature* 284: 452. Mouse Cμ heavy chain immunoglobulin gene segment contains three intervening sequences separating domains.
85. Cattaneo, R., Gorski, J. and Mach, B. (1981) *Nucl. Acids Res.* 9: 2777. Cloning of multiple copies of immunoglobulin variable kappa genes in cosmid vectors.
86. Early, P., Huang, H., Davis, M., Calanc, K. and Hood, L. (1980) *Cell* 19: 981. An immunoglobulin heavy chain variable region is generated from three segments of DNA: V-H, D and J-H.
87. Lenhard-Schuller, R., Hohn, B., Brack, C., Hirama, M. and Tonegawa, S. (1978) *Proc. Nat. Acad. Sci. USA*, 75: 4709. DNA clones containing mouse immunoglobulin K chain genes isolated by *in vitro* packaging into phage λ coats.

88. Mattheyssens, G. and Rabbitts, T. H. (1980) *Proc. Nat. Acad. Sci. USA* 77: 6561. Structure and multiplicity of genes for the human immunoglobulin heavy chain variable region.

89. Nishida, Y., Kataoka, T., Ishida, N., Nakai, S., Kishimoto, T., Bottcher, I. and Honjo, T. (1981) *Proc. Nat. Acad. Sci USA*, 78: 1581. Cloning of mouse immunoglobulin-epsilon gene and its location within the heavy chain gene cluster.

90. Sakano, H., Kurosawa, Y., Weigert, M. and Tonegawa, S. (1981) *Nature* 290: 562. Identification and nucleotide sequence of a diversity DNA segment (D) of immunoglobulin heavy-chain genes.

91. Seidman, J. G. and Leder, P. (1978) *Nature* 276: 790. The arrangement and re-arrangement of antibody genes.

92. Steinmetz, M., Zachau, H. G. and Mach, B. (1979) *Nucl. Acids Res.* 6: 3213. Cloning of immunoglobulin kappa light chain genes from mouse liver and myeloma MOPC 173.

93. Yanawaki-Kataoka, Y., Kataoka, T., Takahashi, N., Obata, M. and Honjo, T. (1980) *Nature* 283: 786. Complete nucleotide sequence of immunoglobulin γ-2b chain gene cloned from a newborn mouse DNA.

94. Zakut, R., Girol, D. and Mory, Y. Y. (1980) *Nucl. Acids Res.* 8: 453. Structure of immunoglobulin γ-2b heavy chain gene cloned from mouse embryo gene library.

95. Bell, G. I., Pictet, R. L., Rutter, W. J., Cordell, B., Tischer, E. and Goodman, H. M. (1980) *Nature* 284: 26. Sequence of the human insulin gene.

96. Cordell, B., Bell, G., Tischer, E., DeNoto, F. M., Ullrich, A., Protet, R., Rutter, W. J. and Goodman, H. M. (1979) *Cell* 18: 533. Isolation and characterisation of a cloned rat insulin gene.

97. Lomedico, P., Rosenthal, N., Efstratiadis, A., Gilbert, W., Kolodner, R. and Tizard, R. (1979) *Cell* 18: 545. The structure and evolution of the two nonallelic rat preproinsulin genes.

98. Perler, F., Efstratiadis, A., Lomedico, P., Gilbert, W., Kolodner, R. and Dodgson, J. (1980) *Cell* 20: 555. The evolution of genes: the chicken preproinsulin gene.

99. Lawn, R. M., Adelman, J., Franke, A. E., Houck, C. M., Gross, M., Najarian, R. and Goeddel, D. V. (1981) *Nucl. Acids Res.* 9: 1045. Human fibroblast interferon gene lacks introns.

100. Nagata, S., Mantei, N. and Weissmann, C. (1980) *Nature* 287: 401. The structure of one of the eight or more distinct chromosomal genes for human interferon-alpha.

101. Tavernier, J., Derynck, R. and Fiers, W. (1981) *Nucl. Acids Res.* 9: 461. Evidence for a unique human fibroblast interferon (IFN-β_1) chromosomal gene, devoid of intervening sequences.
102. Ono, M., Cole, M. D., White, A. T. and Huang, R. C. C. (1980) *Cell* 21: 465. Sequence organisation of cloned intracisternal A particle genes.

L

103. Jensen, E. O., Paludan, K., Hyldig-Nielsen, J. J., Jorgensen, P. and Marcker, K. A. (1981) *Nature* 291: 677. The structure of a chromosmal leghaemoglobin gene from soybean.
104. Sullivan, D., Brisson, N., Goodchild, B., Verma, D. P. S. and Thomas, D. Y. (1981) *Nature* 289: 516. Molecular cloning and organisation of two leghaemoglobin genomic sequences of soybean.
105. Ratzkin, B. and Carbon, J. (1977) *Proc. Nat. Acad. Sci. USA* 74: 487. Functional expression of cloned yeast DNA in *Escherichia coli* (Leu 2).
106. Baldacci, P., Royal, A., Cami, B., Perrin, F., Krvst, A., Garapin, A. and Kourilsky, P. (1979) *Nucl. Acids. Res.* 6: 2667. Isolation of the lysozyme gene of chicken.
107. Lindenmaier, W., Nguyen-Huu, M. C., Lurz, R., Stratmann, M., Blin, M., Wurtz, T., Hauser, H. J., Sippel, A. E. and Schutz, G. (1979) *Proc. Nat. Acad. Sci. USA* 76: 6196. Arrangement of coding and intervening sequences of chicken lysozyme gene.
108. Glanville, N., Durnham, D. M. and Palmiter, R. D. (1981) *Nature* 292: 267. Structure of mouse metallothionein-I gene and its mRNA.
109. Anderson, S., Bankier, A. T., Barell, B. G., de Bruijri, M. H. L., Coulson, A. R., Drouin, J., Eperon, I. C., Nierlich, D. P., Roe, B. A., Sanger, F., Schreier, P. H., Smith, A. J. H., Stagen, R. and Young, I. G. (1981) *Nature* 290: 454. Sequence and organisation of the human mitochondrial genome.
110. Chang, A. C. Y., Lansman, R. A., Clayton, D. A. and Cohen, S. N. (1975) *Cell* 6: 231. Studies of mouse mitochondrial DNA in *Escherichia coli*: Structure and function of the eukaryotic-prokaryotic chimeric plasmids.
111. Drouin, J. (1980) *J. Mol. Biol.* 140: 15. Cloning of human mitochondrial DNA in *E. coli.*
112. Favoeron-Fonty, G., Culard, F., Baldacci, G., Goursot, R., Prunell, A. and Bernardi, G. (1979) *J. Mol. Biol.* 134: 493. The mitochondrial genome of wild-type yeast cells VIII: the spontaneous cytoplasmic "Petite" mutation.
113. Martin, N. C., Miller, D. L. and Donelson, J. E. (1979) *J. Biol.*

Chem. 254: 11729. Cloning of yeast mitochondrial DNA in the *Escherichia coli* plasmid pBR322: identification of tRNA genes.

114. MacLeod, A. R., Karn, J. and Brenner, S. (1981) *Nature* 291: 386. Molecular analysis of the unc-54 myosin heavy-chain of *Caenorhabditis elegans.*

115. Nudel, U., Katcoff, D., Carmen, Y., Zevin-Sonkin, D., Levi, Z., Shaul, Y., Shani, M. and Yaffe, D. (1980) *Nucl. Acids Res.* 8: 2133. Identification of recombinant phages containing sequences from different rat myosin heavy chain genes.

N

116. Hennecke, H. (1981) *Nature* 291: 354. Recombinant plasmid carrying nitrogen fixation genes from *Rhizobium japonicum.*

117. Mazur, B. J., Rice, D., Haselkorn, R. (1980) *Proc. Nat. Acad. Sci. USA* 77: 186. Identification of blue-green algal nitrogen fixation genes by using heterologous DNA hybridisation probes.

O

118. Dugaiczyk, A., Woo, S. L. C., Colbert, D. A., Lai, E. C., Mace, M. L. Jr. and O'Malley, B. W. (1979) *Proc. Nat. Acad. Sci. USA* 76: 2253. The ovalbumin gene: cloning and molecular organisation of the entire natural gene.

119. O'Hare, K., Breathnach, R., Benoist, C. and Chambon, P. (1979) *Nucl. Acids Res.* 7: 321. No more than seven interruptions in the ovalbumin gene: comparison of genomic and double-stranded cDNA sequences.

120. Lindenmaier, W., Nguyen-Huu, M. C., Lurz, R., Blin, N., Stratmann, M., Land, H., Jeep, S., Sippel, A. E. and Schutz, G. (1979) *Nucl. Acids Res.* 7: 1221. Isolation and characterisation of the chicken ovomucoid gene.

P

121. Sun, S. M., Slightom, J. L. and Hall, T. C. (1981) *Nature* 289: 37. Intervening sequences in a plant gene: comparison of the partial sequence of cDNA and genomic DNA of French bean phaseolin.

122. Hitzeman, R. A., Clarke, L. and Carbon, J. (1980) *J. Biol. Chem.* 255: 12073. Isolation and characterisation of the yeast 3-phosphoglycerokinase (PGK) by an immunological screening technique.

123. Chien, Y. H. and Thompson, E. B. (1980) *Proc. Nat. Acad. Sci. USA* 77: 4583. Genomic organisation of rat prolactin and growth hormone genes.

158   K. E. Davies

124. Chang, A. C. Y., Cochet, M. and Cohen, S. N. (1980) *Proc. Nat. Acad. Sci. USA* **77**: 4890. Structural organisation of human genomic DNA encoding the pro-opiomelanocortin peptide.

R

125. Beach, D., Piper, M. and Shall, S. (1980) *Nature* **284**: 185. Isolation of chromosomal origins of replication in yeast.
126. Stinchcomb, D. T. and Davis, R. W. (1979) *Nature* **282**: 39. Isolation and characterisation of a yeast chromosomal replicator.
127. Watanabe, S. and Taylor, J. H. (1980) *Proc. Nat. Acad. Sci. USA* **77**: 5292. Cloning of the origin of replication of *Xenopus laevis*.
128. D'Eustachio, P., Meyuhas, O., Ruddle, F. and Perry, R. P. (1981) *Cell* **24**: 307. Chromosomal distribution of ribosomal protein genes in the mouse.
129. Woolford, J. L. Jr., Hereford, L. M. and Rosbash, M. (1979) *Cell* **18**: 1247. Isolation of cloned DNA sequences containing ribosomal proteins from *Saccaromyces cerevisiae*.
130. Artavanis-Tsakonas, S., Schedl, P., Tschudi, C., Pirrotta, V., Steward, R. and Gehring, W. J. (1977) *Cell* **12**: 1057. The 5S genes of *Drosophila melanogaster*.
131. Brown, D. D., Carroll, D. and Brown, R. D. (1977) *Cell* **12**: 1045. The isolation and characterisation of a second oocyte 5S DNA from *Xenopus laevis*.
132. Clarkson, S. G., Smith, H. O., Schaffner, W., Gross, K. W. and Birnstiel, M. L. (1976) *Nucl. Acids Res.* **3**: 2617. Integration of eukaryotic genes for 5S RNA and histone proteins into a phage lambda receptor.
133. Hershey, N. D., Conrad, S. E., Sodja, A., Yen, P. A., Cohen, M., Davidson, N., Ilgen, C. and Carbon, J. (1977) *Cell* **11**: 585. The sequence arrangement of *Drosophila melanogaster* 5S DNA in recombinant plasmids.
134. Jacq, C., Miller, J. R. and Brownlee, G. G. (1977) *Cell* **12**: 109. A pseudogene structure in 5S DNA of *Xenopus laevis*.
135. Peterson, R. C., Doening, J. L. and Brown, D. D. (1980) *Cell* **20**: 131. Characterisation of two *Xenopus* somatic 5S DNAs and one minor oocyte-specific 5S DNA.
136. Tshudi, C. and Pirrotta, V. (1980) *Nucl. Acids Res.* **8**: 441. Sequence and heterogeneity in the 5S RNA gene cluster of *Drosophila melanogaster*.
137. Arnheim, N. (1979) *Gene* **7**: 83. Characterisation of mouse ribosomal gene fragments purified by molecular cloning.

138. Cockburn, A. F., Newkirk, M. J. and Firtel, R. A. (1976) *Cell* 9: 605. Organisation of the ribosomal RNA genes of *Dictyostelium discoideum*: mapping of the non-transcribed spacer regions.
139. Gerlach, W. L. and Redbrook, J. R. (1979) *Nucl. Acids Res.* 7: 1869. Cloning and characterisation of ribosomal RNA genes from wheat and barley.
140. Glover, D. M. and Hogness, D. S. (1977) *Cell* 10: 167. A novel arrangement of the 18S and 28S sequences in a repeating unit of *Drosophila melanogaster* rDNA.
141. Glover, D. M., White, R. L., Finnegan, J. and Hogness, D. S. (1975) *Cell* 5: 149. Characterisation of six cloned DNAs from *Drosophila melanogaster*, including one that contains the genes for rRNA.
142. Goldsbrough, P. B. and Cullis, C. A. (1981) *Nucl. Acids Res.* 9: 1301. Characterisation of the genes for ribosomal RNA in flax.
143. Kidd, S. J. and Glover, D. M. (1980) *Cell* 19: 103. A DNA segment from *D. melanogaster* which contains five tandemly repeating units homologous to the major rDNA insertion.
144. Philippsen, P., Kramer, R. A. and Davis, R. W. (1978) *J. Mol. Biol.* 123: 371. Cloning of the yeast ribosomal DNA repeat unit in SstI and HindIII lambda vectors using genetic and physical size selections.
145. Wellauer, P. K. and Dawid, I. B. (1977) *Cell* 10: 193. The structural organisation of ribosomal DNA in *Drosophila melanogaster*.
146. Yao, M.-C. and Gall, J. G. (1977) *Cell* 12: 121. A single integrated gene for ribosomal RNA in a eukaryotic, *Tetrahymena pyriformis*.
147. Erion, J. L., Tarnowski, J., Weissbach, H. and Brot, N. (1981) *Proc. Nat. Acad. Sci. USA* 78: 3459. Cloning, mapping and *in vitro* transcription — translation of the gene for the large subunit of ribulose-1, 5-biphosphate carboxylase from spinach chloroplasts.
148. McIntosh, I., Paulsen, C. and Bogorad, L. (1980) *Nature* 288: 556. Chloroplast gene sequence for the large subunit of ribulose bis-phosphate carboxylase of maize.

T

149. Perucho, M., Hanahan, D., Lipsich, L. and Wigler, M. (1980) *Nature* 285: 207. Isolation of the chicken thymidine kinase gene by plasmid rescue.

150. Wilkie, N. M., Clements, J. B., Boll, W., Mantei, N., Lonsdale, D and Weissmann, C. (1979) *Nucl. Acids Res.* 7: 859. Hybrid plasmids containing an active thymidine kinase gene of *Herpes simplex* virus I.

151. Dudler, R., Egg, A. H., Kubli, E., Artavanis-Tsakonas, S., Gehring, W. J., Steward, R. and Schedl, P. (1980) *Nucl. Acids Res.* 8: 2921. Transfer RNA genes of *Drosophila melanogaster.*

152. Dunn, R., Delaney, A. D., Gillan, I. C., Hayashi, S., Tener, G. M., Grigliatti, T., Misra, V., Spur, M. G., Taylor, D. M. and Miller, R. C. Jr. (1979) *Gene* 7: 197. Isolation and character-isation of recombinant DNA plasmids carrying *Drosophila* tRNA genes.

153. Fasiolo, F., Bonnet, J. and Lacroute, F. (1981) *J. Biol. Chem.* 256: 2324. Cloning of yeast methionyl-tRNA-synthetase gene.

154. Gergen, J. P., Loewenberg, J. Y. and Wensink, P. C. (1981) *J. Mol. Biol.* 147: 475. tRNA$_2^{Lys}$ gene clusters in *Drosophila.*

155. Hershey, N. D. and Davidson, N. (1980) *Nucl. Acids Res.* 8: 4819. Two *Drosophila melanogaster* tRNA-gly genes are contained in a direct duplication at chromosomal locus 56F.

156. O'Farrell, P. Z., Cordell, B., Valenzuela, P., Rutter, W. J. and Goodman, H. M. (1978) *Nature* 274: 438. Structure and processing of yeast precursor tRNAs containing intervening sequences.

157. Olah, J. and Feldman, H. (1980) *Nucl. Acids Res.* 8: 1975. Structure of a yeast non-initiating methionine-tRNA gene.

158. Olson, M. V., Langhney, K. and Hall, B. D. (1979) *J. Mol. Biol.* 132: 387. Identification of the yeast DNA sequences that correspond to specific tyrosine-inserting nonsense suppressor loci.

159. Randerath, E., Gupta, R. C., Morris, H. P. and Randerath, K. (1980) *Biochemistry* 19: 3476. Isolation and sequence analysis of two major leucine transfer ribonucleic acids (anticodon Mm-A-A) from a rat tumor, *Morris Hepatoma* 5123D.

160. Sekiya, T., Kuchino, Y. and Nichimura, S. (1981) *Nucl. Acids. Res.* 9: 2239. Mammalian tRNA genes: nucleotide sequences of rat genes for tRNA-asp, tRNA-gly and tRNA-glu.

161. Wittig, B., Wittig, S. and Grunz, H. (1979) *Nucl. Acids Res.* 6: 3759. Cloning of chicken embryo tRNA genes using single stranded nucleosomal DNA highly enriched for tRNA comp-lementary sequences.

162. Lee, D. C., McKnight, S. and Palmiter, R. D. (1980) *J. Biol. Chem.* 255: 1442. The chicken transferrin gene: restriction endonuclease analysis of gene sequences in liver and oviduct DNA.

163. Fried, H. M. and Warner, J. R. (1981) *Proc. Nat. Acad. Sci. USA* **78**: 238. Cloning yeast gene for trichodermin resistance and ribosomal protein L3.
164. Kalfayan, L. and Wensink, P. (1981) *Cell* **24**: 97. Alpha-tubulin genes of *Drosophila*.

U

165. Krol, A., Gallinaro, H., Lazar, E., Jacob, M. and Branlant, C. (1981) *Nucl. Acids Res.* **9**: 769. The nuclear 5S RNAs from chicken, rat and man: U5 RNAs are encoded by multiple genes.
166. Loop, D. R., Kristo, P., Stumph, W. E., Tsai, M. J. and O'Malley, B. W. (1981) *Cell* **23**: 671. Structure and expression of a chicken gene coding for U1-RNA.
167. Guerry-Kopecko, P. and Wickner, R. B. (1980) *J. Bacteriol.* **143**: 1530. Cloning of the URA-1 gene of *Saccharomyces cerevisiae*.

V

168. Riddell, D. C., Higgins, M. J., McMillan, B. J. and White, B. N. (1981) *Nucl. Acids Res.* **9**: 1323. Structural analysis of the three vitellogenin genes in *Drosophila melanogaster*.
169. Wahli, W. and Dawid, I. R. (1980) *Proc. Nat. Acad. Sci. USA* **77**: 1437. Isolation of two closely related vitellogenin genes, including their flanking regions, from a *Xenopus laevis* gene library.
170. Wahli, W., Dawid, I. B., Wyler, T., Weber, R. and Ryffel, G. U. (1980) *Cell* **20**: 107. Comparative analysis of the structural organisation of two closely related vitellogenin genes in *X. laevis*.
171. Widmer, H. J., Jaggi, R. B., Weber, R. and Ryffel, G. U. (1979) *Eur. J. Biochem.* **99**: 23. Enrichment and characterisation of the DNA coding for vitellogenin in *Xenopus laevis*.

Y

172. Barnett, T., Pachl, C. Gergen, J. P. and Wensink, P. C. (1980) *Cell* **21**: 729. The isolation and characterisation of *Drosophila* yolk protein genes.

Misc

173. DeFeo, D., Gonda, M. A., Young, H. A., Chang, E. H., Lowy, D. R., Scolnick, E. M. and Ellis, R. W. (1981) *Proc. Nat. Acad.*

Sci. USA **78**: 3328. Analysis of two divergent rat genomic clones homologous to the transforming gene of Harvey murine sarcoma virus.

174. Henikoff, S. Tatchell, K., Hall, B. D. and Nasmyth, K. A. (1981) *Nature* **289**: 33. Isolation of a gene from *Drosophila* by complementation in yeast.
175. Muskavitch, M. A. T. and Hogness, D. S. (1980) *Proc. Nat. Acad. Sci. USA* **77**: 7362. Molecular analysis of a gene in a developmentally regulated puff of *Drosophila melanogaster.*
176. Robbins, K. C., Devare, S. G., Aaronson, S. A. (1981) *Proc. Nat. Acad. Sci. USA* **78**: 2918. Molecular cloning of integrated simian sarcoma virus—genome organisation of infectious DNA clones.
177. St. John, T. P. and Davis, R. W. (1979) *Cell* **16**: 443. Isolation of galactose-inducible DNA sequences from *Saccharomyces cerevisiae* by differential plaque filter hybridisation.
178. Favera, R. D., Gelmann, E. P., Gallo, R. C. and Wong-Staal, F. (1981) *Nature* **292**: 31. A human onc gene homologous to the transforming gene (v-sis) of simian sarcoma virus.

cDNA clones

A

179. Katcoff, D., Nudel, U., Zevinsonkin, D., Carmon, Y., Shani, M., Lehrach, H., Frischauf, A. M. and Yaffe, D. (1980) *Proc. Nat. Acad. Sci. USA* **77**: 960. Construction of recombinant plasmids containing rat muscle actin and myosin light chain DNA sequences.
180. Kindle, K. L. and Firtel, R. A. (1978) *Cell* **15**: 763. Identification and analysis of *Dictyostelium* actin genes, a family of moderately repeated genes.
181. Merlino, G. T., Water, R. D., Chamberlain, J. P., Jackson, D. A., Elgewely, M. R. and Kleinsmith, L. J. (1980) *Proc. Nat. Acad. Sci. USA* **77**: 765. Cloning of sea urchin actin gene sequences for use in studying the regulation of actin gene transcription.
182. Minty, A. J., Caravatti, M., Robert, B., Cohen, A., Daubas, F., Weyclert, A., Gros, F. and Buckingham, M. E. (1981) *J. Biol. Chem.* **256**, 1008. Mouse actin messenger RNAs: construction and characterisation of a recombinant plasmid molecule containing a complementary DNA transcript of mouse alpha-actin mRNA.
183. Ordahl, C. P., Tilghman, S. M., Ovitt, C., Fornwald, J. and Largen, M. T. (1980) *Nucl. Acids Res.* **8**: 4989. Structure and developmental expression of the chick α-actin gene.

184. Schwartz, R. J., Haren, J. A., Rothblum, K. N. and Dugaiczyk, A. (1980) *Biochemistry* **19**: 5883. Regulation of muscle differentiation: cloning of sequences from alpha-actin messenger ribonucleic acid.

185. Shani, M., Nudel, U., Zevin-Sonkin, D., Zakut, R., Givol, D., Katcoff, D., Carmon, Y., Reiter, J., Frischauf, A. M. and Yaffe, D. (1981) *Nucl. Acids Res.* **9**: 579. Skeletal muscle actin mRNA: characterisation of the 3' untranslated region.

186. Gordon, J. I., Burns, A. T. H., Christmann, J. L. and Reeley, R. G. (1978) *J. Biol. Chem.* **253**: 8629. Cloning of a double stranded cDNA that codes for a portion of chicken preproalbumin.

187. Kiossis, D., Hamilton, R., Hanson, R. W., Tilghman, S. M. and Taylor, J. M. (1979) *Proc. Nat. Acad. Sci. USA* **76**: 4370. Construction and cloning of rat albumin structural gene sequences.

188. Law, S. W. and Dugaiczyk, A. (1981) *Nature* **291**: 201. Homology between the primary structure of alpha-fetoprotein, deduced from a complete cDNA sequence and serum albumin.

189. Liao, W. S. L., Ricca, G. A. and Taylor, J. M. (1981) *Biochemistry*, **20**: 1646. Cloning of rat alpha-fetoprotein 3'-terminal complementary deoxyribonucleic acid sequences and preparation of radioactively labelled hybridisation probes from cloned deoxyribonucleic acid inserts.

190. Lin, Y. and Gross, J. K. (1981) *Proc. Nat. Acad. Sci. USA* **78**: 2825. Molecular cloning and characterisation of winter flounder antifreeze cDNA.

191. Chan, L., Dugaiczyk, A. and Means, A. R. (1980) *Biochemistry* **19**: 5631. Molecular cloning of the gene sequences of a major apoprotein in avian very low density lipoproteins.

C

192. Munjaal, R. P., Chandra, T., Woo, S. L. C., Dedman, J. R. and Means, A. R. (1981) *Proc. Nat. Acad. Sci.* **78**: 2330. A cloned calmodulin structural gene probe is complementary to DNA sequences from diverse species.

193. Richards, D. A., Rodgers, J. R., Supowit, S. C. and Rosen, J. M. (1981) *J. Biol. Chem.* **256**: 526. Construction and preliminary characterisation of the rat casein and alpha-lactalbumin cDNA clones.

194. Griffin-Shea, R., Thireos, G., Kafatos, F. C., Petri, W. H. and Villa-Komaroff, L. (1980) *Cell* **19**: 915. Chorion cDNA clones of *D. melanogaster* and their use in studies of sequence homology and chromosomal location of chorion genes.

195. Jones, C. W., Rosenthal, N., Rodakis, G. C. and Kafatos, F. C. (1979) *Cell* 18: 1317. Evolution of two major chorion multigene families as inferred from cloned cDNA and protein sequences.

196. Spradung, A. C., Digan, M. E., Mahowald, A. P., Scott, M. and Craig, E. A. (1980) *Cell* 19: 905. Two clusters of genes for major chorion proteins of *Drosophila melanogaster*.

197. Tsitilou, S. G., Regier, J. C. and Kafatos, F. C. (1980) *Nucl. Acids Res.* 8: 1987. Selection and sequence analysis of a cDNA clone encoding a known chorion protein of the A family.

198. Frischauf, A. M., Lehrauch, H., Rosner, C. and Bredtker, H. (1978) *Biochemistry* 17: 3243. Procollagen complementary DNA, a probe for messenger RNA purification and the number of type 1 collagen genes.

199. Lehrach, H., Frischauf, A. M., Hanahan, D., Wozney, J., Fuller, F. and Boedtker, H. (1979) *Biochemistry* 18: 3146. Construction and characterisation of pro-alpha 1 collagen complementary deoxyribonucleic acid clones.

200. Myers, J. C., Mon-Li, C., Faro, S. H., Clark, W. J., Prockop, D. J. and Ramirez, F. (1981) *Proc. Nat. Acad. Sci. USA* 78: 3516. Cloning a cDNA for the pro-α2 chain of human type 1 collagen.

201. Yamamoto, T., Sobel, M. E., Adams, S. L., Avvedimento, V. E., DiLauro, R., Pastan, I., de Crumbrugghe, B., Showalter, A., Pesciotta, D., Fietzek, P. and Olsen, B. (1980) *J. Biol. Chem.* 255: 2612. Construction of a recombinant bacterial plasmid containing pro-α1 (I) collagen DNA sequences.

202. Odink, K. G., Fey, G., Webaner, K. and Diggelmann, H. (1981) *J. Biol. Chem.* 256: 1453. Mouse complement components C3 and C4: characterisation of their messenger RNA and molecular cloning of complementary DNA for C3.

203. Cochet, M., Perrin, F., Gannon, F., Krust, A., Chambon, P., McKnight, G. S., Lee, D. C., Mayo, K. E. and Palmiter, R. (1979) *Nucl. Acids Res.* 6: 2435. Cloning of an almost full-length chicken conalbumin double-stranded cDNA.

204. Roberts, J. L., Seeburg, P. H., Shine, J., Herbert, E., Baxter, J. C. and Goodman, H. M. (1979) *Proc. Nat. Acad. Sci. USA* 76: 2153. Corticotropin and beta-endorphin: construction and analysis of recombinant DNA complementary to mRNA for the common precursor.

205. Bhat, S. P. and Piatigorsky, J. (1979) *Proc. Nat. Acad. Sci. USA* 76: 3299. Molecular cloning and partial characterisation of delta-crystallin cDNA sequences in a bacterial plasmid.

206. Negishi, M., Swan, D. C., Enquist, L. W. and Nebert, D. W.

(1981) *Proc. Nat. Acad. Sci. USA*, **78**: 800. Isolation and characterisation of a cloned DNA sequence associated with mouse Ah locus and a 3-methyl cholanthrene-induced form of cytochrome P-450.

D

207. Lewis, J. A., Kurtz, D. T. and Melera, P. W. (1981) *Nucl. Acids Res.* **9**: 1311. Molecular cloning of Chinese hamster dihydrofolate reductase-specific cDNA and the identification of multiple dihydrofolate reductase mRNAs in anti-folate-resistant Chinese hamster lung fibroblasts.

F

208. Kupper, H., Keller, W., Kurz, C., Forss, S., Schaller, H., Franze, R., Strohmaier, K., Marquardt, O., Zaslavsky, V. G. and Hofschneider, P. H. (1981) *Nature* **289**: 555. Cloning of cDNA of major antigen of foot and mouse disease virus and expression in *E. coli*.

G

209. Cummings, I. W., Liu, A. Y. and Saber, W. A. (1978) *Nature* **276**: 418. Identification of a new chicken alpha-globin structural gene by complementary DNA cloning.

210. Fantoni, A., Bozzoni, I., Ullu, E. and Facace, M. G. (1979) *Nucl. Acids Res.* **6**: 3505. Construction of a recombinant bacterial plasmid containing DNA sequences for a mouse embryonic globin chain.

211. Humphries, P., Old, R., Coggins, L. W., McShane, T., Watson, C. and Paul, J. (1978) *Nucl. Acids Res.* **5**: 905. Recombinant plasmids containing *Xenopus laevis* globin structural genes derived from complementary DNA.

212. Kay, R. M., Harris, R., Patient, R. K. and Williams, J. G. (1980) *Nucl. Acids Res.* **8**: 2691. Molecular cloning of cDNA sequences coding for the major alpha- and beta-globin polypeptides of adult *Xenopus laevis*.

213. Liebhaber, S. A., Goosens, M. J. and Kan, Y. W. (1980) *Proc. Nat. Acad Sci. USA* **77**: 7054. Cloning of a complete nucleotide sequence of human 5′-alpha-globin gene.

214. Rabbits, T. H. (1979) *Nature* **260**: 221. Bacterial cloning of plasmids carrying copies of rabbit globin messenger RNA.

215. Richards, R. I., Shine, J., Ullrich, A., Wells, J. R. E. and Goodman, H. M. (1979) *Nucl. Acids Res.* **7**: 1137. Molecular

cloning and sequence analysis of adult chicken beta-globin cDNA.

216. Rougeon, F., Kourilsky, P. and Mach, B. (1975) *Nucl. Acids Res.* 2: 2365. Insertion of a rabbit beta-globin gene sequence into an *E. coli* plasmid.

217. Wilson, J. T., Wilson, J. B., deRiel, J. K., Villa-Komaroff, L., Efstratiadis, A., Forget, B. G. and Weissman, S. M. (1978) *Nucl. Acids Res.* 5: 563. Insertion of synthetic copies of human globin genes into bacterial plasmids.

218. Wood, K. O and Lee, J. C. (1976) *Nucl. Acids Res.* 3: 1961. Integration of synthetic globin genes into an *E. coli* plasmid.

219. Chatterjee, B. and Roy, A. K. (1980) *J. Biol. Chem.* 244: 11607. Messenger RNA for alpha-2μ-globulin of rat liver: purification, partial characterisation of the mRNA and synthesis of a HaeIII restriction fragment as its cDNA probe.

220. Kurtz, D. T. and Nicodemus, C. F. (1981) *Gene* 13: 145. Cloning of alpha-2μ-globulin cDNA using a high efficiency technique for the cloning of trace messenger RNAs.

221. Unterman, R. D., Lynch, K. R., Nakasi, H. L., Dolan, K. P., Hamilton, J. W., Cohn, D. V. and Fiegelson, P. (1981) *Proc. Nat. Acad. Sci. USA* 78: 3478. Cloning and sequence of several $\alpha 2\mu$-globulin cDNAs.

222. Fiddes, J. C. and Goodman, H. M. (1979) *Nature* 281: 351. Isolation, cloning and sequence analysis of the cDNA of the alpha-subunit of human gonadotropin.

223. Fiddes, J. C. and Goodman, H. M. (1980) *Nature* 286: 684. The cDNA for the beta-subunit of human chorionic gonadotropin suggests evolution of a gene by readthrough into the 3'-untranslated region.

224. Goeddel, G. V., Heyneker, H. L., Hozumi, T., Arentzen, R., Hakura, K., Yanswa, D. G., Ross, M. J., Miozzar, G., Grea, R. and Seeburg, P. H. (1979) *Nature* 281: 544. Direct expression in *Escherichia coli* of a DNA sequence coding for human growth hormone.

225. Roskam, W. G. and Rougeon, F. (1979) *Nucl. Acids Res.* 7: 305. Molecular cloning and nucleotide sequence of the human growth hormone structural gene.

226. Seeburg, P. H., Shine, J., Martial, J. A., Baxter, J. D. and Goodman, H. M. (1977) *Nature* 270: 486. Nucleotide sequence and amplification in bacteria of structural gene for rat growth hormone.

227. Keshet, E., Rosner, A., Bernstein, Y., Gorecki, M. and Aviv, H. (1981) *Nucl. Acids Res.* 9: 19. Cloning of bovine growth hormone gene and its expression in bacteria.

H
228. Sleigh, M. J., Both, G. W. and Brownlee, G. G. (1979) *Nucl. Acids Res.* 7: 879. The influenza virus haemmaglutinin gene: cloning and characterisation of a double stranded DNA copy.
229. Schedl, P., Artavanis-Tsakonas, S., Steward, R., Gehring, W. J., Mirault, M.-E., Goldschmidt-Clermont, M., Moran, L. and Tissieres, A. (1978) *Cell* 14: 921. Two hybrid plasmid with *D. melanogaster* DNA sequences complementary to mRNA coding for the major heat shock protein.
230. Bregegere, F., Abastado, J. P., Kvist, S., Rask, L., Lalanne, J. L., Garoff, H., Cami, B., Wiman, K., Larhannmar, D., Peterson, P. A., Gachetin, G., Kourilsky, P. and Dobberstein, B. (1981) *Nature* 292: 78. Structure of C-terminal half of two H-2 antigens from cloned mRNA.
231. Kvist, S., Bregegere, F., Rask, L., Cami, B., Garoff, H., Daniel, F., Wiman, K., Karhammer, D., Abastado, J. P. Gachelin, G., Peterson, P. A., Dobberstein, B. and Kourilsky, P. (1981) *Proc. Nat. Acad. Sci. USA* 78: 2772. cDNA clone coding for part of a mouse H-2 major histocompatibility antigen.
232. Ploegh, H. L., Orr, H. T. and Strominger, J. L. (1980) *Proc. Nat. Acad. Sci. USA* 77: 6081. Molecular cloning of a human histocompatibility antigen cDNA fragment.

I
233. Adams, J. M., Gough, N. M., Webb, E. A., Tyler, B. M., Jackson, J. and Cory, S. (1980) *Biochemistry* 19: 2711. Molecular cloning of mouse immunoglobulin heavy chain messenger ribonucleic acids coding for μ, α, γ-1, γ-2a and γ3 chains.
234. Brown, R. D., Armentrout, R. W., Cochran, M. D. Cappello, J. and Langemeier, S. O. (1981) *Proc. Nat. Acad. Sci. USA* 78: 1755. Construction of recombinant plasmids containing *Xenopus* immunoglobulin heavy chain DNA sequences.
235. Gough, N. M., Webb, E. A. Cory, S. and Adams, J. M. (1980) *Biochemistry* 19: 2702. Molecular cloning of seven mouse immunoglobulin K chain messenger ribonucleic acids.
236. Mushinski, J. F., Blattner, F. R., Owens, J. D., Finkelman, F. D., Kessler, S. W., Fitzmaurice, L., Parker, M. and Tucker, P. W. (1980) *Proc. Nat. Acad. Sci. USA*, 77: 7405. Mouse immuno-globulin-D — construction and characterisation of a cloned sigma-chain cDNA.
237. Obata, M., Yamawaki-Katoaka, Y., Takahashi, N., Katoaka, T., Shimizu, A., Mano, Y., Seidman, J. G., Peterlin, B. M., Leder,

P., and Honjo, T. (1980) *Gene* 9: 87. Immunoglobulin gamma-1 heavy chain genes: structural gene sequences cloned in a bacterial plasmid.

238. Seidman, J. G., Edgell, M. H. and Leder, P. (1978) *Nature* 271: 582. Immunoglobulin light chain structural gene sequences cloned in a bacterial plasmid.

239. Zakut, R., Cohen, J. and Givol, D. (1980) *Nucl. Acids Res.* 8: 3591. Cloning and sequence of the cDNA corresponding to the variable region of immunoglobulin heavy chain MPCII.

240. Bell, G. I., Swain, W. F., Pictet, R., Cordell, B., Goodman, H. M. and Rutter, W. J. (1979) *Nature* 282: 525. Nucleotide sequence of a cDNA clone encoding human preproinsulin.

241. Chan, S. J., Noyes, B. E., Agarwai, K. L. and Skiner, D. F. (1979) *Proc. Nat. Acad. Sci. USA*, 76: 5056. Construction and selection of recombinant plasmids containing full-length complementary DNAs corresponding to rat insulin I and insulin II.

242. Liebscher, D. H., Coutelle, C., Rapoport, T. A., Hahn, V., Rosenthal, S., Prehn, S. and Williamson, R. (1980) *Gene* 9: 233. Cloning of carp preproinsulin cDNA in the bacterial plasmid pBR322.

243. Villa-Kamaroff, L., Efstratiadis, A., Browne, S., Lamedico, P., Tizard, R., Naber, S. P., Chick, W. L. and Gilbert, W. (1978) *Proc. Nat. Acad. Sci. USA* 75: 3727. A bacterial clone synthesising proinsulin.

244. Goeddel, D. V., Leung, D. W., Dull, J. J., Gross, M., Lawn, R. M., McCardliss, R., Seeburg, P. H., Ullrich, A., Yelverton, E. and Gray, P. W. (1981) *Nature* 290: 20. The structure of eight distinct cloned human leukocyte interferon cDNAs.

245. Schwarzstein, M., Streuli, M., Panem, S., Nagata, S. and Weissmann, C. (1980) *Gene* 10: 1. The nucleotide sequence of a cloned human leukocyte interferon cDNA.

246. Tanaguchi, T., Fujiikuriyama, Y. and Muramatsu, M. (1980) *Proc. Nat. Acad. Sci. USA* 77: 4003. Molecular cloning of human interferon cDNA.

247. Tanaguchi, T., Ohno, S., Fujiikuriyama, Y. and Muramatsu, M. (1980) *Gene* 10: 11. The nucleotide sequence of the human fibroblast interferon cDNA.

248. Thang, M. N., Thang, D. C., Chelbialix, M. L., Robertgalliott, B., Commoychevalier, M. J. and Chany, C. (1979) *Proc. Nat. Acad. Sci. USA* 76: 3717. Human leukocyte interferon: relationship between molecular structure and species specificity.

249. Weissenbach, J., Chernajorsky, Y., Zeevi, M., Schulman, L., Soreq, H., Nir, V., Wallach, D., Perricaudet, M., Tiollaus, P.

and Revel, M. (1980) *Proc. Nat. Acad. Sci. USA* **77**: 7152. Two interferon messenger RNAs in human fibroblasts — *in vitro* translation and *Escherichia coli* cloning studies.

250. Yelverton, E., Leung, D. Weck, P., Gray, P. W. and Goeddel, D. V. (1981) *Nucl. Acids Res.* **9**: 731. Bacterial synthesis of a novel human leukocyte interferon.

L

251. Hall, L., Davies, M. G. and Craig, R. K. (1981) *Nucl. Acids Res.* **9**: 65. The construction, identification and characterisation of plasmids containing human alpha-lactalbumin cDNA sequences.

252. Smith, D. F., McClelland, A., White, B. N., Addison, C. F. and Glover, D. M. (1981) *Cell* **23**: 441. The molecular cloning of a dispersed set of developmentally regulated genes which encode the major larval serum protein of *D. melanogaster.*

253. Truelsen, E., Gualing, K., Jochimsen, B., Jorsensen, P. and Marcker, K. A. (1979) *Nucl. Acids. Res.* **6**: 3061. Cloning of soybean leghaemoglobin structural gene sequences synthesised *in vitro.*

254. Wieringa, B., Roskam, W. Arnberg, A., van der Zwaag-Gerritsen, J., Ab, G. and Gruber, M. (1979) *Nucl. Acids Res.* **7**: 2147. Purification of the mRNA for chicken very low density lipoprotein II and molecular cloning of its full length double stranded cDNA.

255. Sippel, A. E., Land, H., Lindenmaier, W., Nguyen-Huu, M. C., Wurtz, T., Timmis, K. N., Giesecke, K. and Schultz, G. (1978) *Nucl. Acids Res.* **5**: 3275. Cloning of chicken lysozyme structural gene sequences synthesised *in vitro.*

M

256. Parnes, J. R., Velan, B., Felsenfeld, A., Ramanathan, L., Ferrini, U., Appella, E. and Seidman, J. G. (1981) *Proc. Nat. Acad. Sci. USA* **78**: 2253. Mouse beta-2-microglobulin cDNA clones: a screening procedure for cDNA clones corresponding to rare mRNAs.

257. Richards, D. A., Blackburn, D. E. and Rosen, J. M. (1981) *J. Biol. Chem.* **256**: 533. Restriction enzyme mapping and heteroduplex analysis of rat milk protein cDNA clones.

258. Arnold, H. H. and Siddiqui, M. A. Q. (1979) *J. Biol. Chem.* **18**: 5641. Cloning of synthetic deoxyribonucleic acid that codes for embryonic cardiac myosin light-chain polypeptide.

259. Umeda, P. K., Sinha, A. M., Jakovcic, S., Merten, S., Hsu, H. J., Subramanian, K. N., Zak, R. and Rabinowitz, M. (1981) *Proc.*

Nat. Acad. Sci. USA **78**: 2843. Molecular cloning of two fast myosin heavy chain cDNAs from chicken embryo skeletal muscle.

O

260. Humphries, P., Cochet, M., Krust, A., Gerlinger, P., Kourilsky, P., Chambon, P. (1977) *Nucl. Acids. Res.* **4**: 2389. Molecular cloning of extensive sequences of the *in vitro* synthesised chicken ovalalbumin structural gene.

261. Monahan, J. J., McReynolds, L. A. and O'Malley, B. W. (1976) *J. Biol. Chem.* **251**: 7355. The ovalbumin gene: *in vitro* enzymatic synthesis and characterisation.

262. Buell, G. N., Wickens, M. P., Carbon, J. and Schimke, R. T. (1979) *J. Biol. Chem.* **254**: 9277. Isolation of recombinant plasmids bearing cDNA to hen ovomucoid and lysosome mRNAs.

263. Stein, J. P., Catterall, J. F., Woo, L. C., Means, A. R. and O'Malley, B. W. (1978) *Biochemistry* **17**: 5763. Molecular cloning of ovomucoid gene sequences from partially purified ovomucoid messenger RNA.

P

264. Gordon, D. F. and Kemper, B. (1980) *Nucl. Acids Res.* **8**: 5669. Synthesis, restriction analysis and molecular cloning of near full length DNA complementary to bovine parathyroid hormone mRNA.

265. Kronenberg, H. M., McDevitt, B. E., Majzoub, J. A., Nathans, J., Sharp, P. A., Potts, J. T. and Rich, A. (1979) *Proc. Nat. Acad. Sci. USA* **76**: 4981. Cloning and nucleotide sequence of DNA coding for bovine preproparathyroid hormone.

266. Cooke, N. E., Coit, D., Shine, J., Baxter, J. D. and Martial, J. A. (1981) *J. Biol. Chem.* **256**: 4007. Human prolactin: cDNA structural analysis and evolutionary comparisons.

267. Gubbins, E. J., Maurer, R. A., Hartley, J. L. and Donelson, J. E. *Nucl. Acids Res.* **6**: 915. Construction and analysis of recombinant DNAs containing a structural gene for rat prolactin cDNA.

268. Nilson, J. H., Thomason, A. R., Horowitz, S., Sasavage, N. L., Blenis, J., Albers, R., Salser, W. and Rottmann, F. M. (1980) *Nucl. Acids Res.* **8**: 1561. Construction and characterisation of a cDNA clone containing a portion of the bovine prolactin sequence.

269. Gedanu, L., Wasnick, M. A., Connor, W., Watson, D. C., Dixon, G. H. and Iatrou, K. (1981) *Nucl. Acids Res.* **9**: 1463. Molecular analysis of the protamine multi-gene family in rainbow trout testis.

270. Jenkins, J. R., Bishop, J. O. and Butterworth, P. H. W. (1979) *Nucl. Acids Res.* 6: 3805. Molecular cloning of three major sequence species from rainbow trout protamine mRNA.

R

271. Hudson, P., Hale, J., Cronk, M., Shine, J. and Niall, H. (1981) *Nature* 291: 127. Molecular cloning and characterisation of cDNA sequences coding for rat relaxin.

272. Bozzoni, I., Beccari, E., Zhong, X. L., Amaldi, F., Pierandrei-Amaldi, P. and Campioni, N. (1981) *Nucl. Acids Res.* 9: 1069. *Xenopus laevis* ribosomal protein genes: isolation of recombinant cDNA clones and study of the genomic organisation.

273. Meyuhas, O. and Perry, R. P. (1980) *Gene* 10: 113. Construction and identification of cDNA clones for mouse ribosomal proteins: application for the study of R-protein gene expression.

S

274. Kistler, M. K., Taylor, R. E., Kandala, J. C. and Kistler, W. S. (1981) *Biochem. Biophys Res. Communs.* 99: 1161. Isolation of recombinant plasmids containing structural gene sequences for rat seminal vesicle secretory proteins IV and V.

275. Seeburg, P. H., Shine, J., Martial, J. A., Ullrich, A., Baxter, J. D. and Goodman, H. M. (1977) *Cell* 12: 157. Nucleotide sequence of part of the gene for human chorionic somatomammotropin: purification of DNA complementary to predominant mRNA species.

276. Shine, J., Seeburg, P. H., Martial, J. A., Baxter, J. D. and Goodman, H. M. (1977) *Nature* 270: 494. Construction and analysis of recombinant DNA for human chorionic somatomammotropin.

277. Geiser, M., Doring, H. P., Wostemeyer, J., Behrens, U., Tillmann, E. and Starlinger, P. (1980) *Nucl. Acids Res.* 8: 6175. A cDNA clone from *Zea mays* endosperm sucrose synthetase mRNA.

278. Pays, E., Delranche, M., Lheureux, M., Vervoort, J., Bloch, J., Gannon, F. and Steinert, M. (1980) *Nucl. Acids Res.* 8: 5965. Cloning and characterisation of DNA sequences complementary to messenger ribonucleic acids coding for the synthesis of two surface antigens of *Trypanosoma brucci*.

T

279. MacGillvray, R. T. A., Friezner Degen, S. J., Chandra, T., Woo, S. L. C. and Davie, E. W. (1980) *Proc. Nat. Acad. Sci. USA* 77:

5153. Cloning and analysis of a cDNA coding for bovine prothrombin.

280. Christophe, D., Porocas, H., Gannon, F., Martynoff, G., Pays, E. and Vassart, G. (1980) *Eur. J. Biochem.* 111: 419. Molecular cloning of bovine thyroglobulin complementary DNA: characterisation of 2,500 base-pair and 1,900-base-pair fragments.

281. Steinmetz, M., Frelinger, J. G., Fisher, D., Hunkapiller, T., Pereira, D., Weissman, S. M., Uehara, H., Nathanson, S. and Hood, L. (1981) *Cell* 24: 125. Three cDNA clones encoding mouse transplantation antigens: homology to immunoglobulin genes.

282. MacLeod, A. R. (1981) *Nucl. Acids Res.* 9: 2675. Construction of bacterial plasmids containing sequences complementary to chicken alpha-tropomysin mRNA.

283. Cleveland, D. W., Lopata, M. A., MacDonald, R. J., Cowan, N. J., Rutter, W. J. and Kirschner, M. W. (1980) *Cell* 20: 95. Number and evolutionary conservation of alpha- and beta-tubulin and cytoplasmic beta- and gamma-actin genes using specific cloned cDNA probes.

284. Sanchez, F., Natzle, J. E., Cleveland, D. W., Kirschner, M. W. and McCarthy, B. J. (1980) *Cell* 22: 845. A dispersed multi-gene family encoding tubulin in *Drosophila melanogaster.*

U

285. Arnemann, J., Heins, B. and Beato, M. (1979) *Eur. J. Biochem.* 99: 361. Synthesis and characterisation of a DNA complementary to pre-uteroglobulin mRNA.

V

286. Cozens, P. J., Cato, A. C. B. and Jost, J.-P. (1980) *Eur. J. Biochem.* 112: 443. Characterisation of cloned complementary DNA covering more than 6,000 nucleotides (97%) of avian vitellogenin mRNA.

287. Ohno, T., Cozens, P. J., Cato, A. C. B. and Jost, J.-P. (1980) *Biochem. Biophys. Acta* 606: 34. Recombinant plasmids containing avian vitellogenin structural gene sequences derived from cDNA.

288. Smith, D. F., Searle, P. F. and Williams, J. G. (1979) *Nucl. Acids Res.* 6: 487. Characterisation of bacterial clones containing DNA sequences derived from *Xenopus laevis* vitellogenin mRNA.

289. Wahli, W., Dawid, I. B., Wyler, T., Jaggi, R. B., Weber, R. and Ryffel, G. U. (1979) *Cell* 16: 535. Vitellogenin in *Xenopus laevis* is encoded in a small family of genes.

Misc

290. Goldstein, E. S. and Arthur, C. G. (1979) *Biochim. Biophys. Acta.* **565**: 265. Isolation and characterisation of cDNA complementary to transient maternal polyA⁺ RNA from the *Drosophila* oocyte.

291. Jacob, E. (1980) *Nucl. Acids Res.* **8**: 1319. Characterisation of cloned cDNA sequences derived from *Xenopus laevis* polyA⁺ oocyte RNA.

292. Mangiarotti, G., Chung, S., Zuker, C. and Lodish, H. F. (1981) *Nucl. Acids Res.* **9**: 947. Selection and analysis of cloned developmentally regulated *Dictyostelium discoideum* genes by hybridisation competition.

293. Mansson, P. E., Sugino, A. and Harris, S. E. (1981) *Nucl. Acids Res.* **9**: 935. Use of a cloned double stranded cDNA coding for a major androgen dependent protein in rat seminal vesicle secretion: the effect of testosterone on gene expression.

294. Ordahl, C. P., Kioussis, D., Tilghman, S., Ovitt, C. E. and Fornwald, J. (1980) *Proc. Nat. Acad. Sci. USA* **77**: 4519. Molecular cloning of developmentally regulated, low abundance messenger RNA sequences from embryonic muscle.

295. Parker, M. G., White, R. and Williams, J. G. (1980) *J. Biol. Chem.* **255**: 6996. Cloning and characterisation of androgen-dependent mRNA from rat ventral prostate.

296. Williams, J. G., Lloyd, M. M. and Devine, J. M. (1980) *Cell* **17**: 903. Characterisation and transcription analysis of a cloned sequence derived from a major developmentally regulated mRNA of *D. discoideum*.